Field and Laboratory Activities in

ENVIRONMENTAL SCIENCE

Fifth Edition

Field and Laboratory Activities in ENVIRONMENTAL SCIENCE

Eldon D. Enger
Delta College

Bradley F. Smith
Western Washington University

with Contributions by Paul Nowak, Jr.

Wm. C. Brown Publishers

ubuque, IA   Bogota   Boston   Buenos Aires   Caracas   Chicago
uilford, CT   London   Madrid   Mexico City   Sydney   Toronto

Book Team

Editor *Margaret J. Kemp*
Production Editor *Karen L. Nickolas*
Designer *Kristyn A. Kalnes*
Photo Editor *Lori Hancock*
Permissions Coordinator *Vicki Krug*
Art Processor *Renee Grevas*

Wm. C. Brown Publishers
A Division of Wm. C. Brown Communications, Inc.

Vice President and General Manager *Bevery Kolz*
Vice President, Publisher *Kevin Kane*
Vice President, Director of Sales and Marketing *Virginia S. Moffat*
Vice President, Director of Production *Colleen A. Yonda*
National Sales Manager *Douglas J. DiNardo*
Marketing Manager *Patrick E. Reidy*
Advertising Manager *Janelle Keeffer*
Production Editorial Manager *Renée Menne*
Publishing Services Manager *Karen J. Slaght*
Royalty/Permissions Manager *Connie Allendorf*

Wm. C. Brown Communications, Inc.

President and Chief Executive Officer *G. Franklin Lewis*
Senior Vice President, Operations *James H. Higby*
Corporate Senior Vice President, President of WCB Manufacturing *Roger Meyer*
Corporate Senior Vice President and Chief Financial Officer *Robert Chesterman*

Copyedited by Toni Good

Cover photo: Ohanapecosh River below Silver Falls, Mount Rainier
National Park, WA © Terry Donnelly/Tom Stack & Assoc.

Some of the laboratory experiments included in this text may be
hazardous if materials are handled improperly or if procedures are
conducted incorrectly. Safety precautions are necessary when you are
working with chemicals, glass test tubes, hot water baths, sharp
instruments, and the like, or for any procedures that generally require
caution. Your school may have set regulations regarding safety
procedures that your instructor will explain to you. Should you have any
problems with materials or procedures, please ask your instructor for
help.

Printed in the United States of America by Wm. C. Brown Communications, Inc.,
2460 Kerper Boulevard, Dubuque, IA 52001

10 9 8 7 6 5 4 3 2 1

Contents

Preface

The exercises within these pages have been prepared to offer the most flexibility in teaching environmental science as it truly is—an interdisciplinary course. The topics in this manual range from investigating ecological principles to conducting campus inventories. Students are encouraged to actively explore environmental concerns—population hazards, toxicity, soil, water and air quality, effects of radiation, impact of human activity on species diversity—and solutions—land-use planning, waste reduction, energy consumption, sustainable agriculture, and influencing public officials. Although some courses take a more regional approach, these exercises are easily adapted to a variety of situations: rural and urban, northern and southern, desert and forested, coastal and inland.

New to this edition are 15 exercises that focus on the "prevention" of environmental hazards and how to be more influential in policy-making (Environmental Law and Decision Making, NGOs—Non-Governmental Organizations). Many exercises are accompanied by assessment sheets, which can be used as hand-in assignments.

Two supplements are available to instructors at no additional cost: *Laboratory Resource Guide* offers lab times, equipment needs, safety precautions, and other helpful hints; *Laboratory Software* offers a quick method for computing chi-square analysis and population data (available in IBM 3.5″ only, ISBN 14176).

We appreciate the comments of Charles Wheeler, Temple Junior College, who provided valuable recommendations for the improvement of this text. We also welcome your comments. If you have suggestions on how the exercises could be improved, or if you would like to see other kinds of topics included, let us know. We may be able to include them in future editions.

Eldon D. Enger
Bradley F. Smith

Field Trip Suggestions

1. Visit a local zoning board meeting or city council meeting when questions of land-use priorities are on the agenda. List the interest groups present at the meeting and describe the major points of view expressed by each group.
2. Visit a park and have a planner show how decisions were made regarding specific land uses. Select one portion of the park and list the reasons for developing it for its particular use.
3. Visit a supermarket and determine the origin of five fruits or vegetables, two types of fresh meats, one canned meat, one canned fish, one package of coffee, one package of tea, and one package of bread. Some of these items will have the origin printed on the label. To find the place of origin of some of the items, you may need to ask the produce manager or the butcher.

    Plot the places of origin on a map of the world. Select one domestic and one foreign item and list the steps necessary to get these items to market.
4. Visit a clothing store and determine the country of origin of five items by reading the labels. Use data from the *United Nations Data Books, The Population Reference Bureau Annual Data Sheet,* or other sources to determine typical wage rates for people in those countries.

    Plot the countries of origin on a map of the world. Write a paragraph describing why you think these items were manufactured where they were.

Alternative Learning Activities

1. Invite a city planner to discuss how decisions are made relative to city growth.
2. Invite a landscape architect to visit the class and discuss factors that are considered in making a plan.
3. Have a speaker discuss the local business climate.
4. Interview local decision makers by asking a series of questions that are important locally. Following this, discuss why different people have different opinions.
5. Visit a shopping mall. Sit and watch. Record the number of different roles (jobs) being performed by the people present. Develop a diagram that shows how these roles interrelate.
6. Trace a polystyrene foam container, glass bottle, or plastic soft-drink bottle from its site of manufacture to its final disposal. (You may need to make phone calls to get the information you need.)
7. Bring in several foreign and domestic foods and drinks. Plot their journey to you on a map of the world. Discuss the importance of these products to the economies of their places of origin. Discuss the roles of politics, energy use, and world hunger as they are related to our use of food.
8. Invite a local corporate leader to discuss corporate policy and environmental issues with the class. Balance this experience by also inviting a local environmental leader to discuss the viewpoint of environmentalists.
9. Conduct library research on several major naturalists and describe how each developed his or her individual philosophy.

Interrelationships Simulation

Objectives

1. Simulate a situation in which different portions of the public have different goals.
2. Allow students to understand that decisions made by political bodies are compromises.
3. Recognize that all the parties in a dispute do not have the same resources.

Safety Considerations: There are no safety considerations in this exercise.

Introduction

In this exercise each student will play a role. It will mirror a situation that is similar to what might occur in the real world: a local zoning board is being asked to make a decision on the possible rezoning of a piece of land from agricultural use to industrial use. The following groups will be represented and will carry out the roles described.

1. The zoning board will hear input from a variety of sources and make a decision about the rezoning of the land. The purpose for the rezoning is to allow the construction of an industrial solvent manufacturing plant just outside the city limits near a major freeway.
2. The industrial developer will present information to the zoning board in favor of the rezoning proposal.
3. Representatives of environmental interests will make presentations opposing the rezoning of the land.
4. Local business leaders will make presentations about how they feel the development will affect them.
5. The media will seek information about the other groups and decide what will be published.

Procedures

1. Students will choose or be assigned a role to play in the simulation.
2. Take time to develop positions and prepare for the simulation. At this time the instructor will provide a sheet of secret information about your group. You will also be given a number of "chips" to use. Any group may elect to buy secret information about any other group at this time.
3. Play out the simulation based on the rules listed.
4. The instructor will lead a discussion following the simulation.
5. Write a paragraph listing three things you learned from the exercise.

Rules

1. Each group will receive a specific number of chips that they can use to obtain time to speak before the zoning commission or to purchase additional secret information from the instructor.
2. Each minute of time before the zoning board will cost one chip.
3. Each item of new information will also cost one chip. To request new information, the group must first pay its chip to the instructor, then specify which group it wants information about. If there is no information, the chip is not returned. If there is information, the instructor will provide one item of information.
4. The secret information that has been purchased does not become public unless it is revealed during the course of speaking before the zoning commission or sold to the media at the price of one item of information per chip.
5. The media cannot buy information from the instructor; they may only buy it from one of the groups.
6. *Each group will be given the additional secret information that pertains to it.*

Zoning Board

Zoning board members will

1. Decide who may speak before the board
2. Listen to the presentations of the various groups
3. Time the presentations
4. Ask questions
5. Make a final decision

The members of the zoning board will be

1.   A real estate agent
2.   A local banker
3.   A housewife
4.   A retired local businessman

Industry

Several members of the class will represent an industry.

1.   *Only one* may speak before the zoning board.
2.   The industry will have a total of twenty chips with which to pay for time before the zoning board and to invest in new information from the instructor.
3.   The basic role of the industry is to convince the zoning board that the rezoning from agricultural to industrial use would be in the best interests of the citizens of the area, because it would create jobs, provide a needed service in a convenient place, and bring additional dollars into the local economy.

Environmentalists

Several members of the class will represent environmental interests.

1.   *Any member of the group* may speak before the zoning board.
2.   They will have five chips with which to buy time before the board or new information from the instructor.
3.   Each member of the group will have at least one chip.
4.   The major role of the environmentalists is to convince the zoning board that the rezoning proposal would be unwise, because it would take extremely productive farmland out of production; increase the runoff from the parking area, causing local flooding; and probably cause air and water pollution. In addition they will argue that the whole project could be built in another location with much less impact.

Local Business Leaders

Several members of the class may represent local business interests.

1.   *Any member of the group* may speak before the zoning board.
2.   They will have ten chips with which to buy time before the zoning board or additional secret information.
3.   Each member of the group will have at least two chips to spend.
4.   The concerns of this group are not very unified. Some members are afraid that the new industry at the edge of town will cause pollution problems that will degrade the quality of life in the community. Others think that the plant will bring new business to the area, and they support it.

The Media

The media will be represented by one or two members of the class.

1.   They will have five chips to spend on secret information.
2.   They can buy secret information only if one of the groups is willing to sell it. One item of information costs one chip. They can buy any information a group is willing to sell, whether it is about the group selling the information or about one of the other groups.
3.   Information that is obtained by the media may be displayed for all to see (written on blackboard or newsprint), but the media are not required to "publish" the information.

Additional Secret Information

The instructor will have secret information that any group except the media may purchase. Each item of information costs one chip.

Interrelationships Simulation
Data Sheet

Name _____

Section _____

In the space below, describe three things that you learned from this simulation.

Scientific Measurements

Objectives

1. Be able to define the following terms: *meter, centimeter, observation, hypothesis, scientific method, precision, metric system, millimeter, gram, kilogram, milligram, liter, milliliter.*
2. Be able to demonstrate an ability to measure length, volume, temperature, and weight using the metric system.

> **Safety Considerations:** Students should exercise care in the use of glassware. If any glass items are broken, they should be disposed of in the appropriate container.
> The thermometers contain mercury. If any thermometer is broken, the mercury should be cleaned up carefully, since it is toxic. Your instructor will be able to provide assistance.

Introduction

The process of science involves the critical evaluation of ideas and information. A scientist has a "healthy skepticism" about information and ideas. A scientist is constantly aware that information must be valid and reliable. In order to be valid and reliable it must repeatedly be shown to be true. The way most scientists evaluate ideas and information has come to be known as the *scientific method.* It is usually thought of as consisting of several kinds of activities. **Observation** is central to any scientific endeavor. The use of the senses to collect information is often extended by the use of mechanical devices, such as microscopes, sensitive microphones, film, or other mechanisms that extend human sensory ability. It is often important to quantify observations in some way. Precise measurements of time, length, volume, temperature, or other variables may be valuable.

A second idea that is important in the scientific method is hypothesis formation. A **hypothesis** is a logical guess that answers a question or explains an observation. A hypothesis is not a fact; it is meant to be tested, challenged, and refined as a result of experience. Therefore, a good hypothesis must have two characteristics: it must account for all of the available data and it must be able to be tested for accuracy.

Once a hypothesis has been formed, it should be checked to see if it is true. This may involve the collection of additional data or the construction of an experimental situation that tests the hypothesis. An experiment usually involves an artificial situation in which all variables are controlled except for the one that the scientist is interested in exploring. Obviously, the collection of information at this point involves accurate observation and precise measurements in the collection of data.

Once a hypothesis has been tested, it is either supported by the new information collected or found not to be valid. If it is not valid, it must be discarded or revised. It is important to understand that just because a hypothesis may turn out to be invalid does not mean that it was not a reasonable way of trying to explain the cause of the original observation.

Another characteristic of the scientific method is that experimental results or observations should be repeatable by the scientist and by others who use the same or similar methods. Most scientists would be very cautious about publishing the results of a single experiment. Only after obtaining the same results repeatedly would they feel comfortable about publishing their findings.

Procedures

During this exercise you will

1. Become familiar with the metric system as it is used to measure length, weight, volume, and temperature.
2. Make observations.
3. Construct hypotheses.
4. Make measurements.
5. Test hypotheses.
6. Evaluate data.

The Metric System of Measurements

Most scientific work uses the metric system of measurements because it is internationally known. (The United States is the only major country of the world that has not officially adopted the metric system of measurement.) The metric system is also much easier to use because it is a decimal system. That means that the various units of measure are related to one another by factors of ten. Following are the most commonly used metric measurements.

Length
Meter
Centimeter = 0.01 meter
Millimeter = 0.001 meter
Kilometer = 1,000 meters

Weight
Gram
Milligram = 0.001 gram
Kilogram = 1,000 grams

Volume
Liter
Milliliter = 0.001 liter = 1 cubic centimeter

Temperature
Water freezes at 0° C and boils at 100° C.
Body temperature is usually about 37° C.
Room temperature is usually between 20° C and 25° C.

Testing Hypotheses

Exercise 2.1

In this exercise you will use the scientific method to clarify the relationship of various characteristics of the human body to one another. For example, you have probably *observed* that there is a relationship between height and weight (tall people usually weigh more than short people), but what is the specific relationship?

You will construct a hypothesis of the relationship between height and weight and test it.

Hypothesis: Weight in kilograms is equal to the height in centimeters divided by 4.

Test the hypothesis by weighing and measuring the height of all the members of the class. Record your data on the data sheet.

Exercise 2.2

You have probably *observed* that tall people have long arms, whereas shorter people have shorter arms. Let us construct a hypothesis.

Hypothesis: The length of a person's arm in centimeters is equal to 0.4 of the person's height in centimeters.

How can this hypothesis be tested?

Collect data. Record data on data sheet.

Exercise 2.3

You have probably *observed* that people differ in the temperature of their hands: some people have warm hands and some people have cold hands. Can you construct a hypothesis concerning hand temperature? Write it in the space provided on the data sheet.

Test your hypothesis by making actual measurements of hand temperature using thermometers marked in ° C. Record your results on the data sheet.

Exercise 2.4

Make an observation and construct a hypothesis about the relationship between the length of the index finger and the volume it takes up. Record information on the data sheet.

Scientific Measurements
Data Sheet

Name _____

Section _____

Exercise 2.1

Student	Height in centimeters	Weight in kilograms	$\frac{\text{Height}}{\text{Weight}}$ (cm/kg)	
			Expected	**Observed**
			4	
			4	
			4	
			4	

1. Is height in centimeters four times the weight in kilograms?

2. What is the average relationship between height and weight?

Exercise 2.2

Student	Height in centimeters	$\times 0.4 =$	Length of arm in centimeters	
			Expected	**Observed**

1. Is the hypothesis correct? Why or why not?

Exercise 2.3

1. Write your hypothesis here.

2. Describe your method of testing the hypothesis.

3. Did you support or disprove your hypothesis?

Exercise 2.4

1. What is your observation?

2. Write your hypothesis here.

3. Describe your method of testing the hypothesis.

4. What did your results show?

Objectives

1. Be able to locate and give the function of each of the following parts of the microscope:
    (1)   ocular lens
    (2)   revolving nosepiece
    (3)   low-power objective lens
    (4)   high-power objective lens
    (5)   stage
    (6)   iris diaphragm
    (7)   lamp
    (8)   coarse adjustment knob
    (9)   fine adjustment knob
2. Be able to make a wet mount of a piece of newspaper and focus one of the letters of a word under the microscope on lower power.
3. Be able to describe objects viewed through the microscope in words or by making a drawing.
4. Be able to use the dichotomous key to identify different kinds of protozoa through the microscope.

Please Note: Microscopes are very expensive to replace; therefore, be particularly careful when handling them.

Use both hands to carry a microscope from the cabinet to your work station.

Hold it upright so that the ocular lens does not slip out.

Use only clean, dry lens paper to clean dust from the glass lenses. Do not use wet paper, paper towels, or other materials that may scratch the lens.

> **Safety Considerations:** Slides and coverslips are glass. Be careful. Do not cut yourself when using them. The coverslips are very thin and easily broken. Dispose of broken glass in the appropriately labeled container.

Introduction

A microscope is the basic instrument for studying things too small to be seen with the unaided eye. There are many different models of microscopes, but they all have certain parts in common. Locate the following parts of the microscope:

Ocular lens: This is also known as the eye piece. It consists of several pieces of glass that have the effect of magnifying objects by a factor of ten. Many ocular lenses have a pointer mounted in them.

Revolving nosepiece: This portion of the microscope allows different objective lenses to be rotated into position.

Low-power objective lens: This lens is located on the nosepiece and has the ability to magnify objects by a factor of ten.

High-power objective lens: This lens is located on the nosepiece and has the ability to magnify objects forty-five times. (Your microscope may have a high-power objective that magnifies by a slightly different amount.)

Stage: This is the flat platform upon which slides are placed.

Stage clips: These structures are used to hold the slide in place on the stage. (*Caution:* some microscopes are fitted with mechanical stages that do not have clips. These models have spring-loaded devices that hold the slide in place while allowing it to be moved about on the stage. Your instructor will show you how to use this mechanism.)

Coarse adjustment knob: This is the large knob at the base of the microscope that varies the distance between the stage and the objective lens. A *small* change in the position of the knob causes a *large* change in the distance between the stage and the lens.

Fine adjustment knob: This is the smaller, inner knob. A *large* change in the position of the knob changes the distance between the objective lens and the stage *very little.*

Iris diaphragm: This part is located below the stage and regulates the amount of light that passes through the opening in the stage. A small lever is used to change the size of the opening through the iris diaphragm.

Lamp: This is located below the stage and supplies the light necessary for viewing objects.

Switch: This is located on the lamp (or near it) and is used for turning the lamp on and off.

Procedures

In this exercise you will examine several objects through the microscope and become familiar with the proper use of the instrument. You will also be introduced to the use of a dichotomous key, which is a commonly used method to help identify different kinds of organisms.

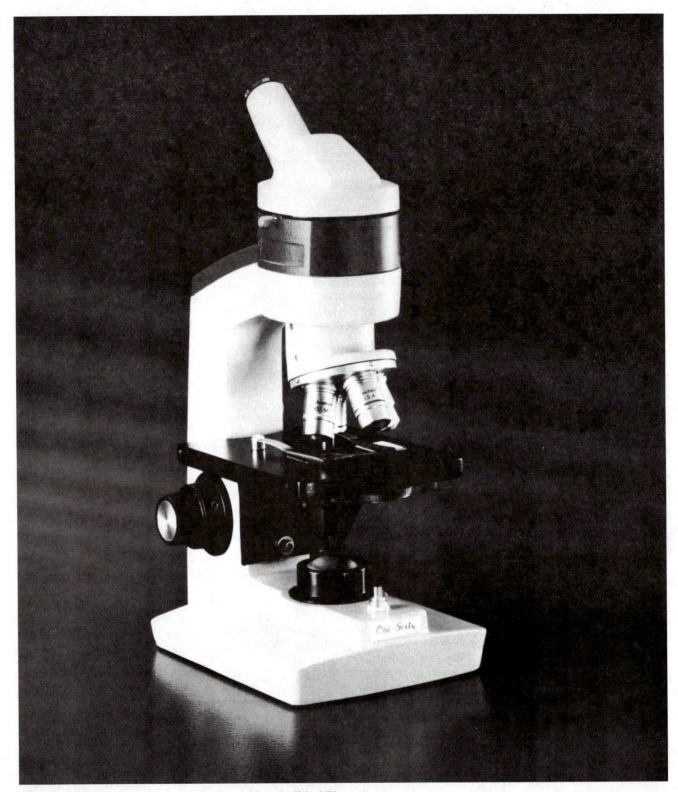

Figure 3.1   Light microscope. Courtesy of Leica, Inc., Buffalo, NY

Magnification

The compound microscope combines the magnifying power of the ocular lens with the magnifying power of the objective lens. The magnifying power of each lens is marked on its tubular housing. Simply multiply the magnification value that is marked on the ocular lens housing by the value marked on the objective lens housing to determine how many times the image of the object has been increased. Notice that your ocular lens magnification is 10×. Since the low-power objective lens is also marked 10×, the total low-power magnification is $10 \times 10 = 100$.

Microscopes often have other objective lenses with other magnifying powers. One of these is the high-power lens. Calculate the total magnifying power of the high-power lens system: _____

Calculate the magnifying power of any other lens systems present: _____

If your microscope has an oil immersion lens, your instructor will give special instructions if you are to use this lens.

Resolving Power

Resolving power is a measure of lens quality. Quality lenses have a high resolving power, which is the capacity to deliver a clear image in fine detail. If a lens has a high magnifying power but a low resolving power, it is of little value. The image may be large but it may not be clear enough to show fine detail.

Clean, unscratched lenses have a better resolving power than dirty, scratched lenses. Therefore, you should clean the lenses regularly, *but use only clean, dry lens paper.*

Field of View

You have already learned that different lenses have different magnifying powers. It is also important to understand that each lens has a specific area that can be observed through it: the higher the magnifying power of an objective lens the smaller the area viewed. This is sometimes hard to appreciate since the size of the circle of light you observe through the ocular lens looks the same. However, when you switch from low power to high power, you are only looking at the central portion of the low-power field of view. Therefore it is important to center an object in the low-power field of view before you switch to high power.

Viewing and Focusing

1. Make a wet-mount slide of a 1 cm × 1 cm piece of newsprint. Use the smallest print available. Place the newsprint on the slide and place two drops of water on top of it. Place a coverslip over the water and newsprint.
2. Center the newsprint over the hole in the stage of your microscope.
3. Rotate the low-power lens into position over the specimen. Make the distance between the lens and the specimen as small as possible by using the coarse adjustment knob.
4. While looking through the ocular lens, use the coarse adjustment knob to move the low-power lens away from the newsprint. You should eventually see some of the fibers in the paper and perhaps a portion of a letter or two.
5. Sharpen the focus by using the fine adjustment knob.
6. Adjust the light by manipulating the iris diaphragm.
7. Center a letter in the low-power field of view.
8. Without adjusting the focus, switch to high power. You should be able to see a portion of your low-power field of view. *Use the fine adjustment knob only to adjust the focus.*
9. If you have difficulty with any of this procedure, ask your instructor for help.

Identification of Protozoa

Make a slide of the live mixed protozoa culture in the lab and use the dichotomous key on page 14 to identify the various kinds of organisms in the mixture. Answer the appropriate questions on the data sheet. This will provide you with practice in the use of the microscope and an introduction to the use of keys that you may be using later in the course.

A dichotomous key is a tool used to help identify the specific kind of organism from within a group. The key presents a series of pairs of statements. You must choose one as being correct. Start at the top of the key with the Roman numeral I. Your two choices are green in color or not green in color. If the organism is green in color proceed to the next statements about swimming ability (A). Biologists use many very detailed keys often using technical language to help identify the many kinds of protozoa. Also you should understand that there are many more kinds of protozoa than those indicated here. This key is a simple example useful in identifying those organisms commonly found in commercially available mixed protozoan cultures.

A Key to the Kinds of Protozoa Typically Found in Mixed Protozoa Cultures

I.   Green in color
    A.   Swims
        1.   Elongate, may change shape—*Euglena* and similar species
        1.   Spherical or egg-shaped, does not change shape
            i.   Tiny, egg-shaped, single cell—*Chlamydomonas* and similar species
            i.   Large sphere of many cells—*Volvox*
    A.   Does not swim; may float or drift—various species of algae (not protozoa)
I.   Not green in color
    A.   Swims
        1.   Has long, whip-like projections at one end; often vibrates as it swims
            i.   Elongates, changes shape—*Peranema* and similar species
            i.   Tiny, spherical or egg-shaped—many kinds of organisms called monads
        1.   Has a large number of tiny, moving, hair-like structures
            i.   Blue in color—*Stentor*
            i.   Colorless
                a.   Attaches with tail end—Rotifer. This is not a protozoan but is often found in cultures.
                a.   Does not attach with tail end
                    2.   Large, slipper-shaped—*Paramecium*
                    2.   Not slipper-shaped—Many other kinds of ciliated protozoa
    A.   Does not swim; body changes shape as it "flows"—*Amoeba*

Microscope
Data Sheet

1. Select one letter from the paper and describe what it looks like through the microscope.

2. How does the image under high power differ from the image under low power?

3. When you move the slide to the right in what direction does the object appear to move under the microscope?

4. Make a wet-mount slide of the protozoa from the culture of mixed protozoa. Observe them through the microscope and identify them using the dichotomous key to the mixed protozoa culture.
   List the protozoa you were able to identify.

   Which was the largest?

   Which swam most rapidly?

Atoms and Molecules

Objectives

1. Define these terms in writing.
   - a. atom
   - b. ion
   - c. isotope
   - d. molecule
   - e. compound
   - f. electron
   - g. proton
   - h. ionic bond
   - i. element
   - j. acid
   - k. base
   - l. covalent bond
   - m. neutron
   - n. orbital
   - o. nucleus

2. Determine the number of protons, neutrons, and electrons for any atom, given the information on a periodic table.
3. Draw and label a diagram of any atom with an atomic number less than 20, given the information on a periodic table.
4. Diagram a conceivable molecule using all the atoms given, and show the shape of that molecule and the proper number of bonds for each atom.
5. Describe in writing and/or by a diagram the characteristics of ionic and covalent bonds.
6. Determine if a given solution is an acid or a base by using any of the following indicators:
   - a. pHydrion paper
   - b. bromthymol blue
   - c. phenolphthalein
   - d. pH meter
7. Interpret the pH number of a solution.
8. Interpret a chemical equation and point out the reactants and the products.

Safety Considerations: Solutions A through E are acids or bases. Be careful not to get these solutions on your hands or clothing because they may be caustic. Disposal of solutions A through E poses environmental hazards; therefore, carefully follow the instructor's directions. Chemicals used as reagents, such as bromthymol blue or sodium iodide, may permanently stain clothing. Use with caution.

Introduction

All matter, including living matter, is made up of small units known as **atoms** and **molecules.** There are ninety-two different kinds of natural atoms, known as **elements.** These atoms can be combined in many ways to form millions of different kinds of molecules. Each kind of atom has a specific arrangement of parts that differs from the arrangement of parts in other kinds of atoms. The specific structure of an atom determines the kinds of atoms that can bond together to form larger molecules. Each atom consists of **protons, neutrons,** and **electrons** arranged with the protons and neutrons in a central area (**nucleus**) and with the electrons moving around in **orbitals** at some distance from the nucleus. The protons are positively charged (+) and the electrons are negatively charged (–); they are equal in number and they are attracted to one another. Since the electrons are light and are moving very rapidly, there is a tendency for the electrons to move around the nucleus. Sometimes an electron may have so much energy that it actually flies away from its nucleus and the atom loses a negative charge. In other cases, atoms may pick up an extra electron and obtain an extra negative charge.

Chart 4.1 is a periodic table of the different kinds of elements that exist. *Do not attempt to memorize it.* Since this table has a lot of information on it, you will need to learn how to interpret this information.

Figure 4.1 interprets the information about a carbon atom from a periodic table. The symbol for carbon is C. The number at the top is the atomic number and tells you how many protons and/or electrons are in the atom. The number at the bottom is the sum of the number of protons and neutrons in the atom. You can determine the number of neutrons in this atom by subtracting the atomic number 6 from the mass number 12; you will find that there are 6 neutrons in a carbon atom.

Preview

This exercise is not designed to make you a chemist. If you are going to understand some of the important biological concepts, it is necessary that you be aware of certain bits of information about atoms and molecules. Work by yourself or in pairs as you proceed through this exercise.

Legend:

$$\begin{array}{l}\text{Atomic Number} \\ \text{(\# of Protons)}\end{array} \longrightarrow \begin{array}{c} 6 \\ \mathbf{C} \\ 12.01 \end{array} \longleftarrow \begin{array}{l}\text{Symbol} \\ \text{Mass Number}\end{array}$$

IA	IIA	IIIB	IVB	VB	VIB	VIIB	VIII			IB	IIB	IIIA	IVA	VA	VIA	VIIA	O	
1 **H** 1.008																	2 **He** 4.003	2
3 **Li** 6.939	4 **Be** 9.012											5 **B** 10.81	6 **C** 12.01	7 **N** 14.01	8 **O** 16.00	9 **F** 19.00	10 **Ne** 20.18	2·8
11 **Na** 22.99	12 **Mg** 24.31											13 **Al** 26.98	14 **Si** 28.09	15 **P** 30.97	16 **S** 32.06	17 **Cl** 35.45	18 **Ar** 39.95	2·8·8
19 **K** 39.10	20 **Ca** 40.08	21 **Sc** 44.96	22 **Ti** 47.90	23 **V** 50.94	24 **Cr** 52.00	25 **Mn** 54.94	26 **Fe** 55.85	27 **Co** 58.93	28 **Ni** 58.71	29 **Cu** 63.54	30 **Zn** 65.37	31 **Ga** 69.72	32 **Ge** 72.59	33 **As** 74.92	34 **Se** 78.96	35 **Br** 79.91	36 **Kr** 83.80	2·8·18·8
37 **Rb** 85.47	38 **Sr** 87.62	39 **Y** 88.91	40 **Zr** 91.22	41 **Nb** 92.91	42 **Mo** 95.94	43 **Tc** (97)	44 **Ru** 101.1	45 **Rh** 102.9	46 **Pd** 106.4	47 **Ag** 107.9	48 **Cd** 112.4	49 **In** 114.8	50 **Sn** 118.7	51 **Sb** 121.8	52 **Te** 127.6	53 **I** 126.9	54 **Xe** 131.3	2·8·18·18·8
55 **Cs** 132.9	56 **Ba** 137.3	57 **La** 138.9	72 **Hf** 178.5	73 **Ta** 180.9	74 **W** 183.9	75 **Re** 186.2	76 **Os** 190.2	77 **Ir** 192.2	78 **Pt** 195.1	79 **Au** 197.0	80 **Hg** 200.6	81 **Tl** 204.4	82 **Pb** 207.2	83 **Bi** 209.0	84 **Po** (210)	85 **At** (210)	86 **Rn** (222)	2·8·18·32·18·8
87 **Fr** (223)	88 **Ra** (226)	89 **Ac** (227)																

Lanthanum Series →

58 **Ce** 140.1	59 **Pr** 140.1	60 **Nd** 144.2	61 **Pm** (147)	62 **Sm** 150.4	63 **Eu** 152.0	64 **Gd** 157.3	65 **Tb** 158.9	66 **Dy** 162.5	67 **Ho** 164.9	68 **Er** 167.3	69 **Tm** 168.9	70 **Yb** 173.0	71 **Lu** 175.0

(71 Lu: 2·8·18·32·9·2)

Actinium Series →

90 **Th** 232.0	91 **Pa** (231)	92 **U** 238.0	93 **Np** (237)	94 **Pu** (242)	95 **Am** (243)	96 **Cm** (247)	97 **Bk** (247)	98 **Cf** (249)	99 **Es** (254)	100 **Fm** (253)	101 **Md** (256)	102 **No** (253)	103 **Lw** (259)

(103 Lw: 32·?·9·2)

Mass numbers of the most stable known isotopes are shown in parentheses.

1—Hydrogen (H)
2—Helium (He)
3—Lithium (Li)
4—Beryllium (Be)
5—Boron (B)
6—Carbon (C)
7—Nitrogen (N)
8—Oxygen (O)
9—Fluorine (F)
10—Neon (Ne)
11—Sodium (Na)
12—Magnesium (Mg)
13—Aluminum (Al)
14—Silicon (Si)
15—Phosphorus (P)
16—Sulfur (S)
17—Chlorine (Cl)
18—Argon (Ar)
19—Potassium (K)
20—Calcium (Ca)
26—Iron (Fe)
29—Copper (Cu)
30—Zinc (Zn)
35—Bromine (Br)
47—Silver (Ag)
53—Iodine (I)
79—Gold (Au)
80—Mercury (Hg)
82—Lead (Pb)
90—Thorium (Th)
92—Uranium (U)

Chart 4.1   Periodic table of elements.

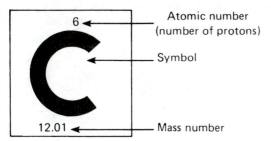

Figure 4.1   Periodic table information.

Caution: This exercise requires the use of glassware and various chemicals; therefore, exercise care when moving around the work area and when handling the materials.

During this lab exercise you will

1.   Diagram the parts of an atom
2.   Assemble molecules that are formed as a result of ionic bonding
3.   Diagram molecules that are formed as a result of covalent bonding
4.   Determine and interpret the pH number of several solutions
5.   Mix reactants and observe the results

Procedure

Review of Atomic and Molecular Structure

The Structure of Atoms

Refer to the periodic table of the elements (chart 4.1). Determine the number of protons and neutrons in the element with the atomic number 7. Write in the numbers in the nucleus illustrated in figure 4.2.

1.   Write the name of the atom and then position the electrons correctly in the orbitals of this atom.

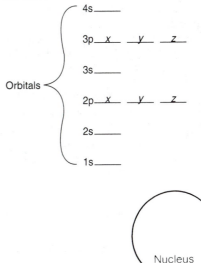

Figure 4.2 _____ .
                       (Name of atom you diagrammed)

2.   Choose any other atom with an atomic number from 1 to 20 on the periodic table of the elements. Determine the name of the element and the number and position of the protons, neutrons, and electrons and indicate this in figure 4.3. If you have difficulty, be sure to ask your instructor for help. You may also want to sketch several other atoms for practice.

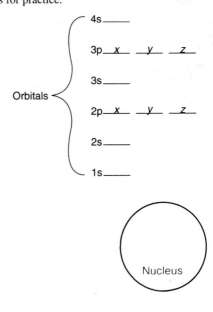

Figure 4.3   _____ .
                (Name of atom you diagrammed)

Molecular Structure

Specific atoms may be combined in certain ways to form larger units called molecules. The bonding together of atoms is a very precise process. Some kinds of atoms are very reactive and will combine with one or two other kinds of atoms. We will not try to determine why certain atoms combine into molecules, but rather how this process happens.

   Ionic Bonds Some kinds of atoms have such a strong attraction for electrons that they will steal from other atoms having electrons that are rather loosely held to the atom. The specific structure of an atom determines whether it will gain or lose electrons to form an ion. All **ions** are formed by either the gain or loss of electrons. Atoms that lose electrons are positively charged (+), and atoms that gain electrons are negatively charged (–). Those ions that have the same charge (both + or both –) will repel one another, whereas those with unlike charges will attract one another and form an **ionic bond** between them. Figure 4.5 has a number of models of different ions. Cut them apart and assemble them to form as many different kinds of **compounds** (combinations of ions) as you can. List at least five of them as figure 4.4. (Chemists generally write the formula of ionic compounds by writing the symbol for the positive ion first and then the negative ion. Then they use a subscript to indicate the number of ions needed to balance the charge.)

Figure 4.4   Ionic compounds.

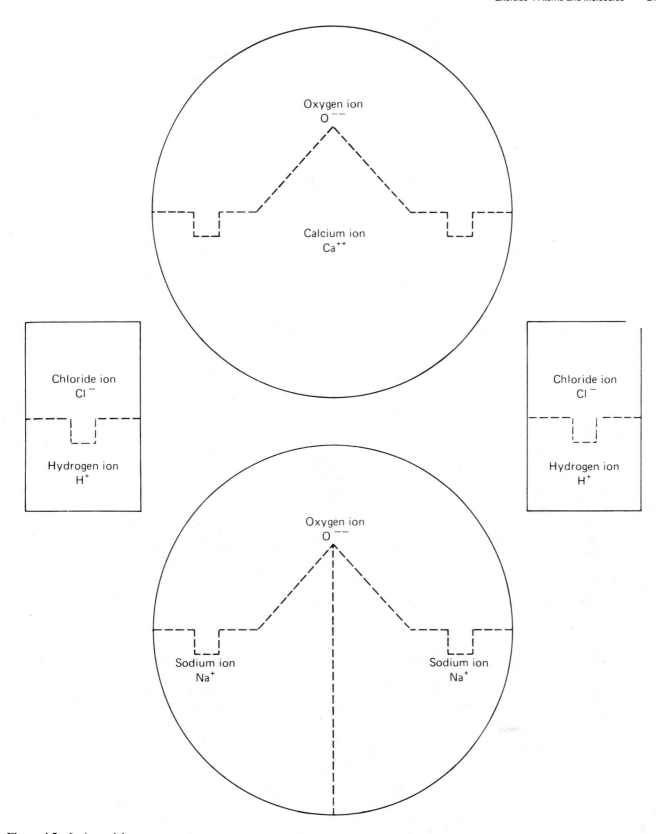

Figure 4.5 Ionic models.

Directions: Refer to the periodic table of elements (chart 4.1) and determine the number of protons and electrons that comprise the following *atoms, ions,* and *molecules.* As you determine the amount of protons or electrons that are a part of a specific ion, the number of +'s indicates how many electrons have been lost; the number of –'s indicates the number of electrons gained.

1.   Protons in a sodium atom (Na) _____   Electrons in a sodium atom (Na) _____
2.   Protons in a sodium ion (Na⁺) _____   Electrons in a sodium ion (Na⁺) _____
3.   Protons in a chlorine atom (Cl) _____   Electrons in a chlorine atom (Cl) _____
4.   Protons in a chloride ion (Cl⁻) _____   Electrons in a chloride ion (Cl⁻) _____
5.   Protons in a sodium   Electrons in a sodium
        chloride molecule (NaCl) _____   chloride molecule (NaCl) _____
6.   Protons in a calcium atom (Ca) _____   Electrons in a calcium atom (Ca) _____
7.   Protons in a calcium ion (Ca⁺⁺) _____   Electrons in a calcium ion (Ca⁺⁺) _____
8.   Protons in an oxygen atom (O) _____   Electrons in an oxygen atom (O) _____
9.   Protons in an oxide ion (O⁻⁻) _____   Electrons in an oxide ion (O⁻⁻) _____
10. Protons in a calcium   Electrons in a calcium
        oxide molecule (CaO) _____   oxide molecule (CaO) _____

Covalent Bond A second kind of bond that holds atoms together to form molecules is known as a **covalent bond.** In these bonds the electrons are not actually transferred from one atom to another as in the formation of ions, but they may be shared by two or more atoms. Each pair of electrons that is shared is the equivalent of one covalent bond. It is possible to diagram molecules by allowing a line to represent a single covalent bond.

Figure 4.6 indicates that a single carbon atom (C) is sharing four electrons with four different hydrogen atoms (H) and that each of the four hydrogen atoms is sharing an electron with the same carbon atom. If you know how many electrons each atom will be able to share, you should be able to diagram a variety of different kinds of molecules. Sometimes two atoms may share more than one pair of electrons, creating a double bond—for example, O = C = O. The diagram of the carbon dioxide molecule indicates that a carbon atom is sharing two electrons with one oxygen atom and two electrons with another oxygen atom. The oxygens are each sharing two electrons with the same carbon atom.

Table 4.1 is a list of a few atoms and the number of electrons they usually share.

Figure 4.6   Covalent bonding.

Table 4.1 Bonding capacity.

Name of atom	Symbol of element	Number of bonds	Bonding capcity
Carbon	C	4	—C—
Nitrogen	N	3	N
Oxygen	O	2	– O –
Hydrogen	H	1	H –

Directions: Use the atoms and the numbers of electrons they can share as listed in table 4.1, and in the space provided, diagram the arrangement of the atoms of the molecules. To do this, begin with the carbon atoms.

1.   Bond these carbon atoms together in a chain or a ring.
2.   Next add the nitrogens if any are called for.
3.   Then add the oxygens if any.
4.   Finally, count the number of electrons in the whole molecule that are still available for bonding. (If this number is equal to the number of hydrogen atoms called for, simply add one hydrogen to each bondable point.) If there are too few hydrogens to complete available bonds, find two free bondable electrons on adjacent atoms and have them form a second bond between themselves (double bond). Now count the available bonding points—the number of hydrogens called for should equal the number of bondable electrons—and simply add the hydrogens where they can share electrons.

Simple Molecules

1. Methane (1 carbon atom and 4 hydrogen atoms): CH_4

2. Ammonia (1 nitrogen atom and 3 hydrogen atoms): NH_3

3. Aldehyde (1 carbon atom, 2 hydrogen atoms, and 1 oxygen atom): CH_2O

4. Water (1 oxygen atom and 2 hydrogen atoms): H_2O

5. Ethane (2 carbon atoms and 6 hydrogen atoms): C_2H_6

More Complex Molecules

6. Ethyl alcohol: C_2H_5OH

7. Nitrogenous compound: $C_2H_5NO_2$

8. Sugar: $C_6H_{12}O_6$

If molecular stick models are available, your instructor will help you demonstrate the three-dimensional arrangement of atoms in the organic compounds listed previously.

Chemical Testing and Reactions

Acids, Bases, and pH

When some materials are dissolved in water, they release hydrogen ions (H^+). Such solutions are known as **acids.** Other materials actually remove hydrogen ions from solutions. These are known as **bases.** It is frequently important to know if a solution is an acid or a base, and a number of methods have been developed to test solutions for their acidity or alkalinity. All of these systems rely on a scale known as the pH scale, which is a measure of the number of hydrogen ions present in a solution. Pure water has both hydrogen (H^+) and hydroxyl ions (OH^-) present in equal numbers; this is called a neutral solution. If there are more H^+ than OH^-, then the solution is an acid. If there are more OH^- than H^+, then it is a base (alkaline). The pH scale shown in figure 4.7 indicates the range of acidity and alkalinity that can exist.

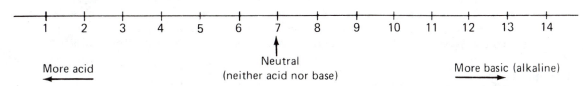

Figure 4.7   pH scale.

Refer to figure 4.7 and answer the following questions.

1. Which of the following pH values indicates the greater concentration of acid?
    pH 1 or pH 4 _____
    pH 5 or pH 3 _____
2. Which of the following pH values indicates the greater concentration of base?
    pH 8 or pH 10 _____
    pH 7 or pH 11 _____
3. What does a pH of 7 indicate? _____

Many kinds of materials change color as the pH of a solution changes. We can make use of this property to determine the pH of unknown solutions. You will find five unknown solutions labeled A through E on the worktable.

1a. Using a pair of forceps, dip a piece of pHydrion paper into solution A.
  b. Compare the color of the pHydrion paper with the information on the paper dispenser chart and record the pH in table 4.2.
2a. Place some of solution A in a test tube and add a few drops of bromthymol blue.
  b. Record the color of the solution in table 4.2.
3a. Place some of solution A in a clean test tube and add a few drops of phenolphthalein.
  b. Record the color of the solution in table 4.2.
4. Repeat these three procedures with solutions B, C, D, and E, and complete table 4.2.

Table 4.2 Unknowns.

Solution	pHydrion paper reading	Color after adding bromthymol blue	Color after adding phenolphthalein
A			
B			
C			
D			
E			

1. Bromthymol blue changes color when mixed with an acid. What color does it become? _____

2. What color is it when in a base? _____

3. What color would you expect it to be if it were in a neutral solution? _____

4. What is the advantage of using pHydrion paper rather than bromthymol blue? _____

_____

pH Meter

Measuring pH using the methods and materials already described has several drawbacks: (1) the solutions to be tested may be destroyed by the addition of indicator chemicals, (2) the accurate determination of a color change is difficult, and (3) a precise determination of pH is almost impossible. The use of an electronic pH metering machine is a better method of determining pH. The pH metering machine can be thought of as a battery that can develop different electrical potentials depending on the hydrogen ion concentration of the solution in which it is placed. The solution serves as a source of the electrical potentials because of its ionic concentration. Even though these electrical potentials are relatively weak, the extreme sensitivity of this machine is able to detect them and display them on a pH scale.

Your instructor will demonstrate the use of a pH meter and provide you with five solutions to be tested. List these solutions on table 4.3 and, following the instructions provided, determine the pH of each. How does this method compare to those used previously?

Table 4.3 pH reading.

Solution	pH meter reading
A	
B	
C	
D	
E	

Reactions

A chemical reaction has occurred when molecules react with one another in such a way that chemical bonds are broken and new molecular combinations are formed as new bonds are formed. In many cases there is physical evidence that a reaction has taken place. This evidence might be some visible change such as the production of a gas that bubbles off, a color change, or the development of an insoluble material that settles to the bottom of the container. As reactions occur between two chemicals, the chemical bonds generally rearrange to form more stable and longer-lasting end products. An example of this occurs when sodium chloride is mixed with silver nitrate. The sodium recombines with the nitrate and the silver becomes associated with chloride.

1. Mix the following ingredients and note the physical changes that occur. Complete the equation by writing in the names and the chemical symbols for the end products.

    Sodium chloride   +   Silver nitrate $\longrightarrow$ _____ + _____

    NaCl   + $AgNO_3$ $\longrightarrow$ _____ + _____

2. Mix sodium iodide with lead nitrate. What is the physical evidence that a reaction has taken place?

3. Complete the equation for the reaction below by putting in the names and the chemical symbols for the reactants.

    _____ + _____ $\longrightarrow$ lead iodide + sodium nitrate

    _____ + _____ $\longrightarrow$ PbI_2 + $2NaNO_3$

4. Mix the following ingredients and note the physical changes that occur. Complete the equation by writing in the names and the chemical symbols for the end products that are not listed.

    HCl + $NaHCO_3$ $\longrightarrow$ CO_2 + _____ + _____

Atoms and Molecules
Data Sheet

Name _____

Section _____

1. What are several differences between ionic and covalent bonds?

2. What things can you learn from the periodic chart of the elements?

3. What do you need to know to diagram an atom?

4. What does a single straight line extending from the symbol for an atom represent?

5. What is the difference between –OH and OH$^-$?

Diffusion and Osmosis

Objectives

1. Define these terms in writing.
   a. solution
   b. solute
   c. solvent
   d. diffusion
   e. relative concentration
   f. differentially permeable membrane
   g. osmosis
   h. net direction of movement
   i. dynamic equilibrium
2. Explain why a particular material diffuses in a particular direction.
3. Determine the net direction of diffusion.
4. Differentiate between diffusion and osmosis.
5. Describe the influence of temperature on the rate of osmosis.
6. Describe the influence varying the concentration of solute and solvent has on the rate of osmosis.

Safety Considerations: Ammonia is a very caustic base. It is dangerous to inhale this gas. Be careful when using the Bunsen burner. An open flame may be a fire hazard. Be especially careful of your hair. Laboratory thermometers do not need to be shaken down; they automatically register higher or lower temperatures. If you accidentally break a thermometer, dispose of the broken glass and mercury as indicated by the instructor.

Introduction

Although you may not know what **diffusion** is, you have experienced this process. Can you remember walking into the front door of your home and smelling a pleasant aroma coming from the kitchen? It was diffusion of molecules from the kitchen to the front of the house that allowed you to detect the odors. To better understand how diffusion works let's consider some information about molecular activity. The kinetic molecular theory states that all materials are made up of molecules, that these molecules have spaces between them, and that they are constantly in motion.

The fact that all things are made up of molecules is easy to accept. Objects must be made of something and if science chooses to call them molecules, it should be OK with us. The second part of the kinetic molecular theory, that there are spaces between molecules, can be demonstrated. If you carefully mix one liter of water with one liter of alcohol, you would expect to have two liters of mixture. Actually you would have a little less than two liters. You know that neither water nor alcohol has been lost because you were very careful in your mixing. One logical explanation for the decrease in volume is that some of the molecules have come to occupy the spaces between the other molecules. Therefore the mixture does not take up as much space as expected.

However, the third aspect of the kinetic molecular theory, the movement of molecules, is the one we want to examine in this exercise. Molecular movement is what is responsible for the process of diffusion. The movement of each molecule is random. That is, it does not have a specific direction in which it should move. The direction of motion of a molecule is likely to be the result of a collision between it and something else. When such a collision occurs, the molecule responds in a way similar to that of a billiard ball as it collides with other billiard balls or the side of the table. The angle between the path of the ball and the side of the table is equal to the angle of the side of the table and the path of the ball as it caroms off. The direction of the movement of a molecule is strictly a matter of physical principle.

The speed of an individual molecule is a measure of its energy. If you add energy to the molecule, it moves faster; if you remove energy, the molecule slows.

Diffusion deals with movement of a group of molecules. We can ask ourselves what will be the overall or average amount and direction of movement of the group of molecules. If ten molecules are all moving randomly within a container, some will hit the right side and bounce off to the left; some will hit the left side and bounce off to the right. If there is no other influence, we would expect that the number hitting one side will be equal to the number hitting the other (figure 5.1a).

A situation in which there is unequal distribution and therefore unequal collisions causes diffusion. If you place all ten molecules in the left half of the container and let them move randomly, you would expect more collision with the left side because that is where all the molecules are located (figure 5.1b). Soon, however, some of the ten molecules, which would have caromed off the left side, would be approaching the right side. The group of molecules is tending to scatter in all directions, due to the random motion of the individual molecules.

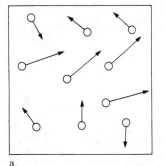 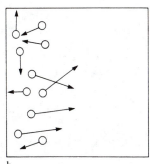

a.   b.

Figure 5.1 Molecular movement.

Diffusion is defined as the net movement of a kind of molecule from an area of higher concentration of that kind of molecule to an area of lower concentration of that kind of molecule.

If you sum the directions of all the molecules toward the right and subtract the directions of all the molecules toward the left, you would have the amount of resultant or net movement. That would be the amount of diffusion. Diffusion is a characteristic of a group of molecules but is understood by dealing with the movement of each molecule.

The net direction of diffusion is always from where there were originally more molecules to where there were fewer. This is similar to the scattering of a crowd of people leaving a theater. Many of the individuals are moving from the theater to the outside, but some are going back to retrieve their gloves or popcorn. However, the net movement is the movement of the individuals leaving the theater minus the movement of those returning.

When the molecules in a container have become equally scattered, the movement to the right becomes equal to the movement to the left, then diffusion has stopped, because there is no net effect. Molecular movement, however, continues. We term this a state of **dynamic equilibrium.**

Preview

The molecular theory states that all substances are made up of molecules that occupy space and are constantly in motion. This exercise will help you examine some phenomena related to this motion of molecules.

During this lab exercise you will

1.   Observe the diffusion of a gaseous material through air
2.   Set up a demonstration of osmosis under a variety of temperature conditions and determine how temperatures influence the rate of osmosis
3.   Graph the results of the osmosis demonstrations

Procedure

Your instructor will open a bottle of ammonia in a corner of the room. Note the time when the bottle is first opened and then the time when you first smell ammonia. The bottle has the highest concentration of ammonia molecules in the room; the ammonia molecules move from this area of highest concentration to where they are less concentrated. Although you cannot actually see this happening, ammonia molecules are leaving the bottle and moving throughout the air in the room because of molecular movement.

If you compare the relative number of ammonia molecules in the bottle to those dispersed in the room, you are dealing with what is called **relative concentration.** Relative concentration compares the amount of a substance in two locations. Whenever there is a difference in concentrations of substance, you can predict in which direction most of the molecules will move. In the case of our demonstration, the ammonia molecules move from the area of higher concentration to the region of lower concentration.

When the bottle is first opened, there are no ammonia molecules in the air, so all the movement of ammonia molecules is from the bottle to the air. Soon, however, the molecules of ammonia mix with the air molecules in the room. Since the ammonia molecules are moving randomly, some of them move from the air back into the bottle. As long as there is a higher concentration of ammonia molecules in the bottle, more of them move out of the bottle than move in. One way of dealing with the direction of movement is to compare the number of molecules leaving the bottle with the number reentering the bottle. This is called the net amount of movement. The movement in one direction minus the movement in the opposite direction is the **net direction of movement.** If, for example, 100 molecules of ammonia leave the bottle and 10 reenter during that time, the net movement is 90 molecules leaving the bottle. Ultimately the number of ammonia molecules moving out of the bottle will equal the number of ammonia molecules moving into it. When this point is reached, the ammonia molecules are said to have reached dynamic equilibrium. It is dynamic because the molecules are still moving, but because this motion is equal in both directions (into and out of the bottle) it is in equilibrium.

The arrow in figure 5.2 indicates the net direction of ammonia movement. The net movement of molecules from the area of higher concentration of ammonia molecules to the area of lower concentration of ammonia molecules is diffusion.

When several kinds of molecules are present, consider only one case of diffusion at a time even though several types of molecules are moving.

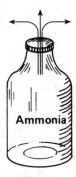

Figure 5.2   Diffusion of ammonia.

In the lungs, a series of tubes transports gases. These tubes become very small and eventually end at a series of small alveolar sacs. Adjacent to these sacs are a number of blood capillaries. By the process of diffusion, oxygen and carbon dioxide are exchanged between the alveolar sacs and the blood in the capillaries.

1.   Draw a heavy arrow on figure 5.3 to indicate the net direction of carbon dioxide movement.
2.   Draw a dotted arrow on figure 5.3 to indicate the net direction of oxygen movement.

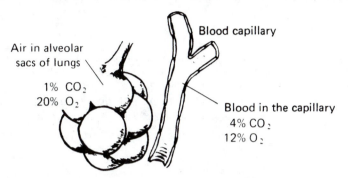

Figure 5.3   Diffusion of oxygen and carbon dioxide.

3.   On figure 5.4, draw an arrow to show the net direction of sugar movement.

Figure 5.4   Sugar diffusion.

4.   Figure 5.5 shows a **differentially permeable membrane.** This is a thin sheet of material that selectively allows certain molecules to cross it but prevents others from crossing. The membrane in this figure is permeable only to water molecules.

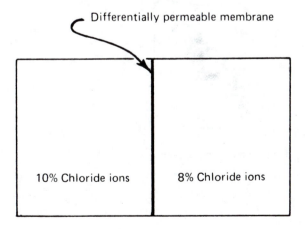

Figure 5.5   Diffusion of water.

What percentage of water is in the left part of the container? _____

What percentage of water is in the right part of the container? _____

Where is the water in higher concentration, left or right side? _____

Draw an arrow to indicate the net direction of movement of the water molecules.

In each of the previous examples, the net movement was a result of diffusion of molecules from a place of higher concentration to a place of lower concentration. The rate at which diffusion occurs is related to the amount of energy the molecules have. Adding energy doesn't change relative concentrations, nor does it influence the direction of diffusion; it merely affects the rate at which diffusion occurs.

Osmosis

Osmosis is a specific type of diffusion; it is the movement of a solvent through a differentially permeable membrane. Biologically we are concerned with water solutions. When water is mixed with other molecules, this mixture is called a **solution.** Water is the **solvent** and the dissolved substance is the **solute.** A solution is characterized by the solute; for example, water and molasses would be characterized as a molasses solution. In osmosis we are not interested in what the solute is, but only in the solvent (water), which can move across the membrane. The example in figure 5.5 was diffusion of water from one side of a membrane to the other; therefore, it was also an example of osmosis.

Osmosis Demonstration

Working in groups prepare three sacs to demonstrate osmosis. Obtain three pieces of dialysis tubing (sausage casing) and soak them in tap water for about one minute. Shake off the excess water and, with a piece of string, securely tie one end of each to form three tubular bags. Fill the tubular bags with full-strength molasses. Leave a small amount of space in the bags. Tie the open ends of each and rinse molasses from the outside of each bag.

Your instructor will tell you if the rate of diffusion into the bags is to be measured by judging the firmness of the bags or by weighing them. If the bags are to be weighed, mark the horizontal line on the graph on the data sheet with the weight of the lightest bag and make a suitable scale in grams along the vertical line. If the bags are not to be weighed, notice the appearance of the bag and gently feel the extent of firmness.

Place one bag in a beaker of tap water and record the temperature. Place a second bag in water heated to 60° C. Place the third bag in a beaker of ice water and record the temperature. Make certain that each bag is completely covered with water (figure 5.6). After five minutes, remove each bag and gently squeeze each to assess any changes in firmness, or weigh each. Note the size, shape, firmness, and weight of each bag. Record your data and return each bag to its appropriate container of water.

Repeat your measurements at five-minute intervals (measurements are to be taken at zero, five, ten, fifteen, and twenty minutes). Using a dashed line for the hot water, solid line for room temperature water, and dotted line for the ice water, record the results of your measurements on the graph on the data sheet.

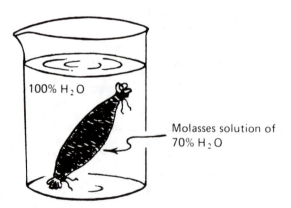

Figure 5.6   Osmosis model.

Diffusion and Osmosis
Data Sheet

Name _____

Section _____

1. In what direction are water molecules capable of moving? Consider molecular movement rather than net movement to answer this.

2. Specify the net direction of water movement by marking an arrow on figure 5.6.

3. In a perfectly tied and unbroken bag, do you see any evidence of sugar molecules passing through the "membrane"? Qualify your answer in terms of how differential permeability operates.

4. From the data in figure 5.7, what is the influence of temperature on the rate of osmosis?

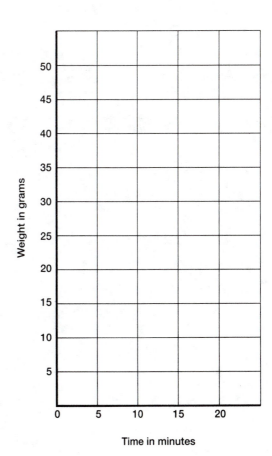

Figure 5.7

5.   Was equilibrium reached in any of the molasses demonstrations?

6.   Why does a good cook wait until just before serving to put salad dressing on a salad? What happens to a lettuce leaf in terms of osmosis when it is "dressed"?

7.   The human cell contains 0.9% solutes (dissolved materials). There is 99.1% water in these cells. The Pacific Ocean contains 3.56% salt. Although this seems like a silly question, how much water is in the ocean? You are cast adrift in this ocean. What would happen to your cells if you were to drink the saltwater?

Field Trip Suggestions

1. Visit local habitats—grassland, desert, forest, or other locally available habitats. Collect five plants that you can take back to the lab to identify.

    Collect five invertebrate animals. Record in detail where each was found. Upon returning to the laboratory, identify the animals.

2. Visit various aquatic systems, such as a stream, a lake, a bog, an irrigation canal, or another locally available aquatic system.

    Collect five plants that grow in or adjacent to the water that you can take back to the lab to identify.

    Collect five invertebrate animals. Record in detail where each was found. Upon returning to the laboratory, identify the animals.

3. Visit various controlled ecosystems, such as a sewage treatment plant, an agricultural field, a municipal park, a forestry plantation, or another locally available area.

    Describe five ways in which each controlled ecosystem is different from a similar natural one.

    Describe five organisms that are aided by the human control of ecosystems and five that are harmed.

4. Visit a farm and record five examples of good soil-conservation practices and five examples of poor soil-conservation practices.

Alternative Learning Activities

1. Observe the food preferences of different species of birds at bird feeders. What types of foods are eaten? Does seed size make a difference? Does the kind of feeder make a difference? Does the size of the perch make a difference?
2. Track an animal in the mud or snow. Identify the animal and try to determine what it was doing from the tracks you find.
3. Place some soil in a broth made by boiling hay and observe the changes in protozoa present over a two-week period.
4. Observe birds at a bird feeder. Which species are the dominant species that chase other birds away? If you can identify individuals, can you determine which individuals within a species are the dominant ones?
5. Dissect a recently living frog and look for parasites in the lungs, body cavity, bladder, and digestive system.
6. Describe the stages of succession you can observe along the edge of a pond, in an abandoned field, or in a sidewalk crack.
7. Photograph several stages of succession you can observe in your neighborhood.
8. Invite a pro-hunting group, such as Ducks Unlimited, or a local wildlife professional to discuss hunting and its value to nongame species.
9. Participate in a local habitat modification project aimed at increasing the numbers of certain kinds of organisms.
10. Research the history of Planned Parenthood and the contribution of Margaret Sanger to the movement. Write a paper on the history of reproductive enlightenment.
11. Culture fruitflies for one month in a closed container. Take a census of the population on a regular basis, prepare a graph of the results, and discuss the outcome of the project.
12. Survey a community and ask questions about desirable family size and actual family size. Correlate your data with the age of the respondent.
13. Prepare a graph showing the population growth of your local urban community over the past twenty years. Select five changes (political, economic, sociological, housing, etc.) in the community that appear to be related to population change and describe how they are related.
14. Dig a hole or observe a road cut and identify the different layers of a soil profile.

Ecosystem Interdependence and Collapse

Objectives

1. Understand how ecosystems work and how ecosystem elements interrelate.
2. Experience the potential consequences of eliminating an active element of an ecosystem.
3. Understand the role humans play in changing or influencing ecosystems.

Introduction

Ecosystem interrelatedness is a cornerstone of environmental science. Interrelatedness means that all things, living and nonliving, are connected to, interdependent on, and related in some way to one another. This exercise helps to take a look at interrelatedness in a simple ecosystem model which the class will construct.

Procedures

This exercise begins with each individual class member assuming the role of one component or element of an ecosystem. Living components include carnivores, herbivores, plants, algae, bacteria, and fungi. Nonliving components include climate, lakes, rivers, oceans, mountains, and plains. Take some time in class to brainstorm and list as many components of your ecosystem as you can. You do not have to have a lot of specific species, but make sure humans are included in your list.

Once you have settled on your ecosystem components, arrange yourselves in a circle starting with the most fundamental components (i.e., sun, water and/or land, and air). Use pieces of twine to connect each ecosystem component to other components on which that first component depends. These links will continue until each classmate is strung together in a web of life. Note which way the dependencies flow. This web represents a simple model of an ecosystem. Note how many components of the model ecosystem humans are dependent on.

Now list potential ecosystem threats that may affect one or more of the components of your ecosystem. These threats could be, for example, changing climate, increased toxins in the environment, or changes in species activities.

Choose one threat from your list. We will use that threat to eliminate one component of your ecosystem (note: at the start, do not remove land, air, or water). When removing an ecosystem component, have that person give a tug on the strings he or she is holding while being removed from the web. Have each person who received a tug give a lesser tug back on the strings they are holding. See how many different threats are needed before everyone has received at least one tug.

Note what effects the removal of one component has on all the other components of your ecosystem. Are other components worse off? Are some components better off? Were some components of the ecosystem eliminated as a result of other components being removed? Repeat the elimination of components several times. Note the degradation of the web as each additional component is removed.

Ecosystem Interdependence and Collapse Data Sheet

Name _____

Section _____

Your instructor will now divide the class into two groups to discuss the following questions.

1. Now that you are able to better visualize all the factors in our ecosystem, is it easier to understand why the human impacts on ecosystems are so important and so potentially dangerous to humans and other species? List some human impacts on the environment and discuss ways these impacts may affect ecosystem health.

2. If humans are impacting natural ecosystems, do you think this problem is likely to increase or decrease in the future? What national or global trends do you think contribute to the changing impact of humans on the environment?

3. In our web of life, we felt tugs on our strings when one component of the ecosystem ceased to function. What types of situations, events, or circumstances do you think these tugs represent?

4. Of the list of five most feasible reasons for removing a component from your ecosystem, how many of these reasons are strongly induced by human activity? Are there clear steps to be taken to lessen this human impact?

Soil Formation and Erosion Processes

Objectives

1. Undertake several simple experiments to facilitate your understanding of processes involved in soil formation.
2. Understand the long-term natural processes involved in soil formation and soil loss including the role of wind, water, and vegetation.
3. Describe possible ways to reduce soil erosion in agricultural settings.

Safety Considerations: Use safety glasses when heating pieces of rock. Use tongs to transfer hot objects.

Introduction

Soil formation and soil erosion are the two opposing forces that determine the amount of soil in a given area. In about two-thirds of the world, topsoil erosion is outpacing the formation of new soil. One of the reasons soil erosion is a problem is that natural soil formation processes are very slow. It takes about 500 years to accumulate 2–3 centimeters of new soil while soil erosion, even at a moderate rate, can remove the same amount of topsoil in 20 years.

Procedures

1. Select two pieces of limestone or fine sandstone. Vigorously rub the pieces together over a cloth. (If you do not have access to natural stones, a brick or concrete block will work equally well.) Observe the amount of time needed to accumulate even a small amount of loose material.
2. This step requires the use of a Bunsen burner or a hot plate (use extreme caution and safety equipment when doing this exercise). In a heat-safe dish, place a small (approximately 50 grams) piece of limestone over the Bunsen burner or hot plate. Heat the limestone for 3–5 minutes. When the limestone is hot, use tongs to drop it into a container of ice water. Note your observations. Were they what you expected?
3. Completely fill a small glass jar with water and screw the lid on tightly. Place the jar in an empty coffee can. Let the jar freeze (not solidly) outside, or in the freezer if you live in a warmer climate. Note what happens to the jar. What conclusions can you draw concerning the freeze–thaw cycle and its contribution to soil formation?
4. Place small pieces of limestone (approximately 1 cm in diameter) in a jar with 200–300 ml of vinegar. Place the uncovered jar over a Bunsen burner or on a hot plate and observe as the contents are heated. What forms on the stones? What do you think causes this to occur? What type of change is this: physical or chemical?
5. As you clean up the lab area, note how easy it is to remove the fine grains of sand from the cloth. A simple brushing action over a waste can will remove most of the particles. Shaking out the cloth outside will remove the vast majority of the remaining particles. If you wanted to get the cloth perfectly particle-free, you might wash it in a washing machine. What conclusion can you draw about your cleanup activities and soil erosion? Was it easier and faster than the soil formation processes in steps 1–4? Can you envision natural analogies to your cleanup process?

Interpretation

Although the results of most of these experiments were fairly easy to predict, did you draw a linkage between temperature changes or other physical processes and soil formation? Did you draw linkages between chemical processes and soil formation?

Rubbing two rocks together generates fine-grained particles. If you continued rubbing the rocks together, you could easily produce a spoonful of particles in a matter of minutes. In the natural environment, this process of rocks rubbing and scraping occurs constantly, in glaciers and moving water, and with wind. Over thousands of years, tremendous quantities of rock particles of all sizes are generated by these natural processes.

In addition, temperature transformations contribute to soil formation. In many regions, especially arid regions, the sun warms the rocks during the day causing expansion. During the night, heat leaves the rock by radiation and conduction, reducing the temperature of the rock in the process. This continual heating and cooling cycle produces small and large fractures in the rock. You observed this type of fracturing when you dropped the hot limestone into the container of ice water.

Similarly, the freeze/thaw cycle contributes to soil formation. When water freezes, it expands with tremendous force. If the water in your jar had been allowed to freeze long enough, it would have broken your jar. In the natural environment, water seeps into cracks in the rock and freezes, shattering the rock into small pieces. This freeze/thaw cycle is particularly hard on paved roads and contributes greatly to road maintenance and repair costs in much of North America.

The bubbles that formed on the limestone pieces when placed in the vinegar were carbon dioxide gas. This gas formed when the limestone [$Ca(CO_3)_2$] reacted with the acid vinegar. Carbon dioxide gas dissolves in moist soil, forming a weak carbonic acid. This acid reacts just as the vinegar did to cause breakdown of limestone and marble. The dissolving effect of weak carbonic water is several times greater than that of pure water. Since much of the limestone actually dissolves into the liquid solution and is carried away, it takes a great amount of limestone (several meters) to make even a few centimeters of soil.

From your cleanup activities, we can see that soil, or small particles of rock in our case, are easily transported and moved from one location to another. These particles are easily moved by mechanical means including wind and water. Because soil is so easily transported, it can erode much more quickly than it is formed. What types of forces in the environment must soil be protected from if it is going to remain in place and not erode?

Through these simple experiments, you probably have a sound understanding of the fragility of soil in our modern society. Industrialized agriculture, which often involves tilling the soil several times per year, can lead to unacceptable annual losses of topsoil. Agricultural practices that minimize or eliminate soil loss will become more common in the years ahead.

Soil Characteristics and Plant Growth

Objectives

1. Observe differences in radish plant growth in different kinds of soil.
2. Measure biomass production of radish plants following two weeks of growth in different soils.

Soil Characteristics

Soil consists of a mixture of different-sized mineral materials, air, moisture, and living organisms. In this exercise we will look at the effect of particle size on water-holding capacity, pore space for air penetration, and plant growth. Plants need three resources from the soil in order to grow. The roots need oxygen to carry on aerobic respiration and grow. Water is needed as a raw material for photosynthesis. Mineral nutrients, such as nitrogen, phosphorous, and potassium, as well as many minor nutrients are needed for the addition of new living material to the plant. In this exercise we will look at five different soils: pea gravel; sand; clay; commercial potting soil; and a mixture of 1/3 gravel, 1/3 sand, and 1/3 clay.

Procedure

Exercise 8.1 The effect of soil moisture on growth

1. Obtain five 8 cm (3-inch) pots and fill each to within 1 cm of the top with one of the five soil mixtures available.
2. Plant five sprouting radish seeds in each pot 1 cm below the surface.
3. Pour 100 ml of water into the pot and catch and measure the amount of water that passes through the pot. You may need to pour very slowly for some of the soils. Record the amount of water passing through on the data sheet at the end of exercise 8.2.
4. Add 25 ml of water to each pot each day for two weeks.
5. At the end of two weeks record the number of plants growing in each pot. Record this information on the data sheet.
6. At the end of two weeks remove the plants from the soil and weigh the total amount of biomass produced. Record data on the data sheet.
7. Measure the length of the roots and record average root length on the data sheet.

Exercise 8.2 The effect of soil nutrients on growth

1. Obtain five 8 cm (3-inch) pots and fill each to within 1 cm of the top with one of the five soil mixtures available.
2. Plant five sprouting radish seeds in each pot 1 cm below the surface.
3. Pour 100 ml of a commercially available 20-20-20 plant food solution into the pot and catch and measure the amount of solution that passes through the pot. You may need to pour slowly for some of the soils. Record the amount passing through on the data sheet at the end of this exercise.
4. Add 25 ml of regular water to each pot each day for two weeks.
5. At the end of two weeks remove the plants from the soil and weigh the total amount of biomass produced. Record data on the data sheet.
6. Measure the length of the roots and record average root length on the data sheet.
7. Compare your data with those from exercise 8.1 to see if the plant food had any effect on the growth of the seedlings.

Exercise 8.3

1. Examine a small amount of sand, clay, and potting soil under a microscope.
2. Note differences in size and shape of soil particles.

Soil Characteristics and Plant Growth Data Sheet

Name _____

Section _____

Exercise 8.1

Kind of soil	Water held (ml)	Biomass produced (grams)	Root length (mm)
Pea gravel			
Sand			
Clay			
1/3 gravel, sand, clay			
Potting soil			

1. Which soil showed the greatest biomass production? What soil characteristics do you think contributed to these results and why?

2. Which soil showed the least biomass production? Why?

3. Do the sizes of the root masses differ in different soils? How?

Exercise 8.2

Kind of soil	ml plant food solution held	Biomass produced (grams)	Root length (mm)
Pea gravel			
Sand			
Clay			
1/3 gravel, sand, clay			
Potting soil			

Compare your results with those from exercise 8.1. Did the presence of plant food affect the biomass produced? How might the size of the soil particles or the nature of the soil particles influence the amount of nutrients held in the soil?

Abiotic Influences on Organisms

Objectives

1. Define these terms in writing.
   a. habitat
   b. control
   c. variable
   d. gradient
2. Demonstrate how to conduct a habitat preference test designed to isolate environmental variables.
3. Collect and interpret data in an attempt to determine which environmental variables are significant.
4. Describe the function of a control in an experimental procedure.

Safety Considerations: Handle hypodermic syringes with care. Do not fill them and squirt their contents anywhere. They are not to leave the lab due to possible contamination if used illegally. Heat lamp bulbs are designed to get very hot. Be careful when moving them. Flood lamps may also get hot.

Introduction

The individuals within a population of a given organism respond to certain features of their environment and ignore others. Many characteristics of a **habitat** (the space an organism inhabits) are **variable** from time to time or at different locations within the habitat. Temperature, quantity of light, and pH often vary in aquatic habitats. When a specific environmental factor varies continuously over a distance, a **gradient** exists. Light intensities can range from absolute darkness to extreme brightness. A shady spot may be a few degrees cooler than a position in direct sunlight only a few meters away. The pH of a lake or stream may also vary from place to place. It seems logical to expect that certain conditions would be most suitable for an organism to thrive and therefore be best for the population. Organisms relate to other organisms as well.

If we are to attempt to determine the relative significance of an environmental condition, we need to isolate it from other variables. Then we can present an organism with several alternative conditions along the gradient. If the organisms respond differently at different positions along the gradient, we can see that the particular variable is significant to the organism.

In this exercise you must apply the scientific method. You will set up an experimental situation, carefully and accurately collect data, and finally analyze the data you collect to determine which environmental variables are significant to the organism and how great the differences must be before the organism responds.

Preview

During this exercise you will work in an assigned group. Each group will work with one variable, such as light, pH, or temperature, and will determine how the organisms (brine shrimp) respond.

1. Each group places brine shrimp in the apparatus.
2. Adjust your apparatus so that you establish the specific environmental gradient assigned to you.
3. Allow the brine shrimp sufficient time to move to their preferred position along the gradient.
4. Collect data concerning population density at five positions along the gradient.
5. Report your data to the class.
6. Record data collected by other groups.
7. Interpret all the data reported.

Since this exercise utilizes the scientific method and experimentation, your instructor may begin the session with a brief discussion to help generate ideas. You could consider some possible techniques for collecting data. Consider including the following ideas in this discussion session.

How should the organism be removed from the apparatus without distorting the results of the experiment?
Should the organisms be flushed from each area or should each area be emptied with no rinse? If you rinse, how much rinse should you apply to each section?
Should you count every individual or should you sample your populations from each section?
Should you count only live organisms or both living and dead organisms?

Procedure

The class will be divided into five groups. Groups will be organized as: group 1 (control), group 2 (pH), group 3 (temperature), group 4 (light), and group 5 (combination of variables).

Each group will need to obtain a test container. (Your instructor may have a specific piece of tubing or a plastic trough for your group to use.) Place the brine shrimp into the test container. Apply the specific variable and allow the container to remain undisturbed for thirty minutes.

Each group must use the same technique throughout the procedure. Any group failing to be consistent with the procedure will invalidate the results.

Group Specifics

Group 1 (Control)

You are working with no variables and no gradients. Take care that all conditions are identical along the length of the container. Leave the container undisturbed for thirty minutes. This container will be used as a basis of comparison by the other groups, i.e., a **control.**

Group 2 (pH)

You are working with one variable, pH. Slowly inject 0.5 ml of 1% HCl into the container at one end. Next, inject 1 ml of 1% KOH at the other end. Leave the container undisturbed for thirty minutes.

Group 3 (Temperature)

You are working with one variable, temperature. Cradle the left end of the container in a plastic bag of crushed ice. Place an infrared heat lamp 30 cm above the other end. Leave the container undisturbed for thirty minutes. Check to make sure the container is horizontal.

Group 4 (Light)

You are working with one variable, light. Place a bright lamp at least 1.5 m from one end of the container. It must be far enough away so that you don't heat up this end of the container and accidentally set up a temperature gradient. Place a vertical cardboard partition at 50 cm. Leave the container undisturbed for thirty minutes.

Group 5 (Combination of Variables)

You are working with a combination of two variables, which you choose from those used by groups 2, 3, and 4. Read the instructions for the specific groups to help determine your setup. After filling the container, leave it undisturbed for thirty minutes.

Data Gathering

After allowing thirty minutes for your brine shrimp to respond to the environmental gradient you established,

1.   Divide the container into five sections, sealing off each one.
2.   Empty the contents of each section into a separate beaker.
3.   Label your beakers so you can identify which beaker came from each section of the container.
4.   Record the pH and temperature of each beaker.
5.   Count the individuals by following the decision reached during the class discussion, or by using the method indicated by your instructor.
6.   Report your data and remain until all groups have reported. You must record data from all other groups because your data, by itself, is worthless. Record all data on the data sheet.
7.   Interpret the data collected and try to determine the habitat preferences of brine shrimp.

Abiotic Influences on Organisms
Data Sheet

Name _____

Section _____

Section I	Section II	Section III	Section IV	Section V

0 cm   20 cm   40 cm   60 cm   80 cm   100 cm

Groups	I	II	III	IV	V	Total Population
1 Control	pH _____ Temp. _____ Pop. _____ % _____	pH _____ Temp. _____ Pop. _____ % _____	pH _____ Temp. _____ Pop. _____ % _____	pH _____ Temp. _____ Pop. _____ % _____	pH _____ Temp. _____ Pop. _____ % _____	_____
2 pH	**Acid** pH _____ Temp. _____ Pop. _____ % _____	pH _____ Temp. _____ Pop. _____ % _____	pH _____ Temp. _____ Pop. _____ % _____	pH _____ Temp. _____ Pop. _____ % _____	**Base** pH _____ Temp. _____ Pop. _____ % _____	_____
3 Temp.	**Cold** pH _____ Temp. _____ Pop. _____ % _____	pH _____ Temp. _____ Pop. _____ % _____	pH _____ Temp. _____ Pop. _____ % _____	pH _____ Temp. _____ Pop. _____ % _____	**Hot** pH _____ Temp. _____ Pop. _____ % _____	_____
4 Light	**Dark** pH _____ Temp. _____ Pop. _____ % _____	pH _____ Temp. _____ Pop. _____ % _____	pH _____ Temp. _____ Pop. _____ % _____	pH _____ Temp. _____ Pop. _____ % _____	**Light** pH _____ Temp. _____ Pop. _____ % _____	_____
5 Combination	pH _____ Temp. _____ Pop. _____ % _____	pH _____ Temp. _____ Pop. _____ % _____	pH _____ Temp. _____ Pop. _____ % _____	pH _____ Temp. _____ Pop. _____ % _____	pH _____ Temp. _____ Pop. _____ % _____	_____

$$\% = \frac{\text{number of individuals in section}}{\text{total number of individuals in all sections}}$$

1.   What was the purpose for having the control?

2.   What was the purpose for having the multiple variables for group 5?

3. Why is it necessary to have large numbers of organisms?

4. How would you modify your procedures if you were to repeat this exercise.

5. Using your interpretation of the data, what do you think are the preferences of this organism in terms of pH, temperature, and light?

Successional Changes in Vegetation

Objectives

1. Define these terms in writing.
   a. succession
   b. pioneer species
   c. climax community
   d. ecosystem
   e. niche
2. List five differences between an old field and a temperature deciduous forest.
3. Describe five changes that occur to the plant life as an area changes from an old field community to a temperature deciduous forest community.
4. State several ways in which the plant life of an area influences the animal life of that area.

Introduction

The analysis of communities of organisms is a rather difficult task. One should be able to identify all the kinds of plants and animals, describe the relationships among different organisms, and measure the sizes of populations. We are not that expert, so we will use artificial designations to identify types of plants and not worry about their true names. In this exercise, we will need to estimate the size of some populations but not try to characterize all the kinds of organism interactions.

The various plant and animal populations interacting in an area receive their energy from the sun and constitute an **ecosystem.** Each particular organism has a **niche**—a job to perform—in the ecosystem. **Succession** is a predictable pattern of change that occurs within ecosystems. **Pioneer plants** are the first type to become established in a barren area. These plants shade the soil and lower the soil temperature, reduce the wind velocity at the soil surface, add organic material to the soil, and support various types of animals. Thus the presence of the pioneer plants changes the environment of an area. A second group of plant types can become established in the area. These new varieties of plants eliminate the pioneer species. The new varieties, in turn, also change the environment of the area. A third group of plant types then grows and replaces the second group of plants. This constant change in the environment and the resulting changes in the plant and animal populations are succession. Eventually a plant community becomes established that is able to maintain its species; new types do not invade the area. This last stage of succession is the climax stage. It may require several hundred years to undergo succession from pioneer plants to the **climax community.**

Procedure

Your instructor will have stretched out 100 m of string in a line so that approximately 50 m is in one stage of succession and 50 m is in a different stage of succession.

This string will have a knot every 10 m. These knots denote places where you will collect information (sampling stations). There are eleven sampling stations.

The class will be divided into groups. Each group will have a specific job to do. The groups will collect data and record them on the data sheet provided.

Group 1 (Small Plants)

1. You will need the following equipment:
   a. Wire hoop
   b. Meterstick
   c. Data sheet
   d. Pencil
2. At each of the knots on the string, place the hoop on the ground and count the number of plants in each of the following categories:
   a. Narrow-leaved plants, grass, and so on
   b. Herbs under 50 cm tall
   c. Herbs over 50 cm tall
3. If there are a lot of plants, try to make an accurate estimate.
4. Record this information on your data sheet.

Group 2 (Large Plants)

1. You will need the following equipment:
    a. String, 2 m long with clamp on the end
    b. Pencil
    c. Data sheet
    d. Meterstick
2. At each knot on the string, attach your 2 m length of string. Travel around the knot, which denotes the center of your sample point, and count all the woody plants. If it is branched below 1 m, record it as a shrub. If it branches above 1 m, record it as a tree. If only a portion of a tree falls inside the circle, record the portion that does (i.e., 1/2, 1/8).
3. Record all data on the data sheet provided.

Group 3 (Estimate Average Height of Dominant Vegetation)

1. You will need a meterstick.
2. At each sampling station, estimate the average height of the dominant (most conspicuous) vegetation as indicated in the figure.
3. Record data on the data sheet provided.

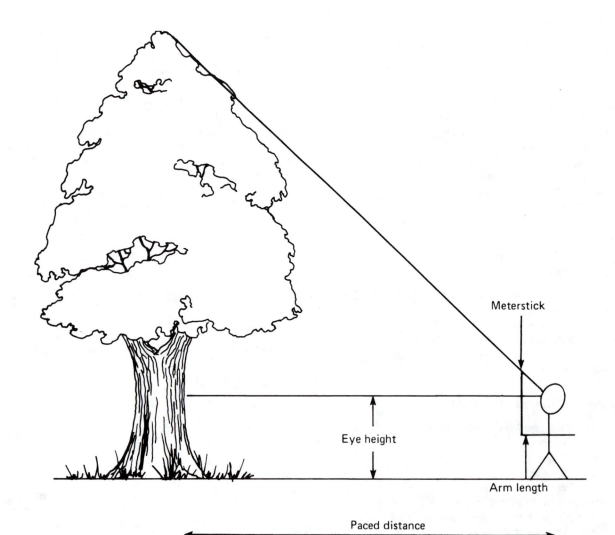

Height of tree = eye height + paced distance

In order to estimate the height of a tree, measure the distance from your eye to your hand held at eye level. Form a right triangle by holding the meterstick vertically with your hand at eye level. The distance from eye to hand and the distance the meterstick protrudes above your hand must be equal. Position yourself so that you can sight over the top of the meterstick to the top of the tree. The height of the tree is equal to the distance from where you are standing to the base of the tree plus the distance from the ground to your eye.

Group 4 (Physical Conditions)

1. You will need the following equipment:
    a. Two soil thermometers
    b. Two regular thermometers
    c. Air speed indicator
    d. Light meter
    e. Relative humidity apparatus
2. At each of the sampling stations, record the following information:
    a. Soil temperature (you must leave the thermometer in the ground for five minutes)
    b. Air temperature (hold the thermometer in such a way that the sun doesn't shine directly on it and the wind doesn't blow directly on it)
    c. Wind speed (hold wind speed indicator 1 m above the ground)
    d. Light reading (hold light meter 1 m above the ground)
    e. Relative humidity (the instructor will explain how to use the apparatus)
3. Record all data on the data sheet.

Successional Changes in Vegetation
Data Sheet

Name _____

Section _____

	Station 1	Station 2	Station 3	Station 4	Station 5	Station 6	Station 7	Station 8	Station 9	Station 10	Station 11
Narrow-leaf plants, e.g., grass											
Herbs under 50 cm tall											
Herbs over 50 cm tall											
Shrubs (branches at 1 m or less)											
Trees (branches above 1 m)											
Average height of dominant vegetation											
Air temperature											
Soil temperature											
Wind speed											
Light meter reading											
Relative humidity											

1. Using the information gathered, describe the climate of the communities you visited.

2. Was there a difference in stages of succession in the various stations where you collected data? At which station was succession at its earliest; at which station was it closest to climax?

3. What human influences did you notice that have interfered with the normal process of succession? What is their effect?

4. Which station do you think showed the greatest variety of organisms and complexity? Why do you think this is true?

Exercise 11   Intraspecific and Interspecific Competition

Objectives

1. Assess the effect of competition among plants at different population densities on the growth of plants.
2. Assess the relative competitiveness of two species of plants when they are planted together.

Competition

Competition is an interaction between organisms in which both are harmed by the interaction. Competition can occur among members of the same species and is called **intraspecific competition** or it can occur between members of different species, in which case it is called **interspecific competition.** Competition of animals can involve fights over food, water, nesting sites, or mates or may involve less overt forms such as greater height allowing some animals access to food that others cannot reach.

Among plants competition usually is difficult to visualize. However, certain resources, such as water, soil nutrients, and sunlight, are in limited supply. Plants that have special abilities to capture these resources should be more successful in competition.

In this exercise we will examine the impact of planting seeds at different densities on the growth of individuals within the population. We will also look at how two kinds of plants influence one another when they are planted together.

Procedure

Exercise 11.1 Intraspecific Competition—Radish

1. Obtain radish seeds that have been soaked so that they are beginning to sprout.
2. Obtain three 8 cm (3-inch) pots and fill to within 2 cm of the top with sand. Press down gently.
3. Carefully place sprouting radish seeds on the sand and cover with about 1 cm of additional sand as follows:
    In pot 1 plant one seed.
    In pot 2 plant ten seeds.
    In pot 3 plant twenty seeds.
4. Pour water into the pot until it begins to run out the bottom.
5. Place in a warm, well-lighted place.
6. Examine daily and water the plants when the surface becomes dry.
7. After three weeks remove the plants from each pot, including roots.
8. Weigh the total biomass produced from each pot, determine the average biomass per plant, and record on the data sheet.

Exercise 11.2 Intraspecific Competition—Wheat

1. Obtain wheat seeds that have been soaked so that they are beginning to sprout.
2. Obtain three 8 cm (3-inch) pots and fill to within 2 cm of the top with sand. Press down gently.
3. Carefully place sprouting wheat seeds on the sand and cover with about 1 cm of additional sand as follows:
    In pot 1 plant one seed.
    In pot 2 plant ten seeds.
    In pot 3 plant twenty seeds.
4. Pour water into the pot until it begins to run out the bottom.
5. Place in a warm, well-lighted place.
6. Examine daily and water the plants when the surface becomes dry.
7. After three weeks remove the plants from each pot, including roots.
8. Weigh the total biomass produced from each pot, determine the average biomass per plant, and record on the data sheet.

Exercise 11.3 Interspecific Competition—Radish and Wheat

1. Obtain radish and wheat seeds that have been soaked so that they are beginning to sprout.
2. Obtain three 8 cm (3-inch) pots and fill to within 2 cm of the top with sand. Press down gently.
3. Carefully place sprouting seeds on the sand and cover with about 1 cm of additional sand as follows:
    In pot 1 plant one radish seed and one wheat seed.

     In pot 2 plant ten radish seeds and ten wheat seeds.

     In pot 3 plant twenty radish seeds and twenty wheat seeds.

4. Pour water into the pot until it begins to run out the bottom.
5. Place in a warm, well-lighted place.
6. Examine daily and water the plants when the surface becomes dry.
7. After three weeks remove the plants from each pot, including roots.
8. Separately weigh the total biomass of radish and wheat plants produced from each pot, determine the average biomass per plant for both the radish and wheat plants, and record on the data sheet.

Intraspecific and Interspecific Competition Data Sheet

Name _____

Section _____

Exercise 11.1 Radish: Intraspecific Competition

Seeds per pot	Total biomass per pot (grams)	Average biomass = $\dfrac{\text{Total biomass}}{\text{Number of seeds per pot}}$
1		
10		
20		

Graph your results below.

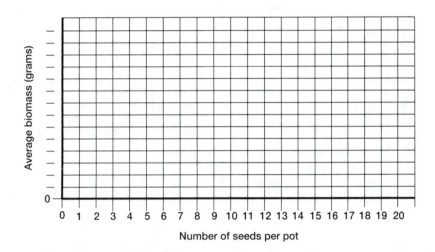

1. At what density was there a change in the biomass per plant as the density of the plants in the pots increased?

2. How did it change?

3. What resources might have been limited?

Exercise 11.2 Wheat: Intraspecific Competition

Seeds per pot	Total biomass per pot (grams)	Average biomass = $\dfrac{\text{Total biomass}}{\text{Number of seeds per pot}}$
1		
10		
20		

Graph your results below.

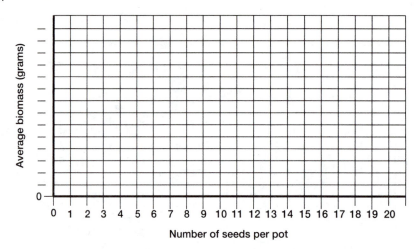

1. At what density was there a change in the biomass per plant as the density of the plants in the pots increased?

2. How did it change?

3. What resources might have been limited?

Exercise 11.3 Interspecific Competition (Radish and Wheat)

Seeds per pot	Total biomass per pot (grams)	Average biomass = $\dfrac{\text{Total biomass}}{\text{Number of seeds per pot}}$
1 radish		
1 wheat		
10 radish		
10 wheat		
20 radish		
20 wheat		

Graph your results below. Use different colors to show the two different kinds of plants, and label the curves.

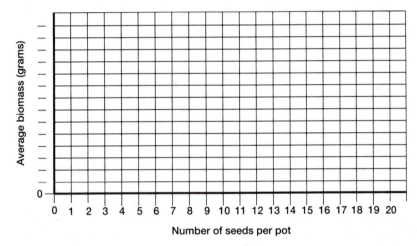

1.   Which kind of plant was most affected by the competition between the two species? In order to answer this question you may want to compare your data with those of exercises 11.1 and 11.2. An easy way to do this would be to include those data on this graph with two other colors. Label the curves.

2.   List three reasons one kind of plant might have been more successful than the other. (How might one of the plants have had a competitive advantage?)

Objectives

1. Compare several differences between polluted and unpolluted streams.
2. Learn techniques for evaluating the quality of water in a stream.

Introduction

The kinds of plants and animals that can live in a stream are determined by several factors, such as the following:

1. Sediment that enters the stream from the land the stream drains
2. Organic compounds that enter with sewage, falling leaves, and land runoff in the drainage basin
3. Oxygen concentration of the water
4. Temperature of the water
5. Fertilizers
6. Toxic materials that enter from the air or from runoff
7. Bacterial levels

This exercise compares two kinds of streams: one that is reasonably clean and has most of its water from uninhabited areas and one that is reasonably dirty and receives runoff from residential and agricultural areas. This will necessitate a field trip in order to visit both types of streams.

Procedure

The class will be divided into groups, with each group having a specific task to accomplish. Each group will have equipment to use to gather data for its part of the exercise. Please follow the directions provided in this exercise or with the specialized equipment you will be using.

I.   Physical Character of the Stream (Group I)

   A.   Temperature
   Use a thermometer to take a temperature reading of the stream. This reading should be taken in a free-flowing portion of the stream. In some situations you may need to attach the thermometer to a pole or wade into the stream. Record results.

   B.   pH
   The pH of the stream can be measured using a portable pH meter. Take measurements in a free-flowing portion of the stream. Record results.

   C.   Total Suspended Matter
   Use a standard container, such as a 1-liter flask, to capture water from a free-flowing portion of the stream. This can be stoppered and returned to the lab to determine the amount of suspended solids in the following manner.

   1.   Obtain a glass fiber filter disk from the desiccator and weigh it very carefully (to the milligram). Record the weight.
   2.   Assemble the filter apparatus as directed by your instructor.
   3.   Thoroughly mix the water in your sample to resuspend any solids that might have settled to the bottom.
   4.   Filter 100 ml of the sample.
   5.   Carefully place the filter on a stainless or aluminum dish and dry in an oven at 103–105° C for one hour.
   6.   At the end of one hour cool the filter in a desiccator. Then weigh it. Record weight.
   7.   Repeat the drying and weighing cycle until the weight doesn't change.
   8.   Calculate the amount of suspended solids by subtracting the initial weight of the clean filter from the final weight of the "dirty" filter in milligrams per liter.

$$\text{Total suspended matter (mg/l)} = \frac{(\text{weight of ``dirty'' filter}) - (\text{weight of clean filter}) \times 1{,}000}{\text{size of sample filtered in ml}}$$

II.   Chemical Nature of the Stream

These exercises make use of kits that allow you to measure the levels of certain chemicals present in the water. Please follow the directions furnished with the kits.

    A.   Oxygen Determination (Group II)

The amount of oxygen in the water is critical for the organisms that live in it. Several factors influence oxygen level. (i) The organic matter entering a stream from runoff, sewage, and other sources results in lower oxygen concentrations, because the breakdown of organic matter to CO_2 and H_2O is an oxygen-requiring process. (ii) The amount of plant life in the streams can also raise O_2 levels when the plants are photosynthesizing or lower it as they decay. (iii) Finally, the amount of turbulence (tumbling over rocks, etc.) can increase the amount of O_2 able to dissolve in the water. Carefully follow the directions furnished with the kit. Repeat the test until you obtain consistent results. Record results.

    B.   Total Phosphate Determination (Group III)

Phosphates may enter streams from several sources. (i) Fertilizers are a common source of phosphates in agricultural and urban areas, as runoff from lawns and farmland contributes phosphates. (ii) Many insecticides are phosphate-containing compounds, which may show up in streams as a result of agricultural practices. (iii) Other human uses, such as detergents, may also add phosphates to a stream. Carefully follow the directions furnished with the kit and determine the amounts of orthophosphate, metaphosphate, and organic phosphates present. Repeat each test until you obtain consistent results. Record results.

    C.   Nitrogen Compounds (Group IV)

Nitrogen compounds (ammonia [NH_3], nitrate [$-NO_3$], and nitrite [$-NO_2$]) may enter streams from several sources. Runoff from pasture lands or feed-lots can add significant ammonia. Likewise septic systems that drain into streams can add ammonia. Nitrate and nitrite can also be added from the atmosphere (automobile exhaust contains a variety of nitrogen oxides) or as fertilizer runoff from land. Follow the directions furnished with the nitrogen kits and determine the amount of ammonia, nitrate, and nitrite in the water. Repeat each test until you obtain consistent results. Record results.

III.   Biological Nature of the Stream

    A.   Bottom Organisms (Group V)

The organisms in a stream are often a good indicator of the quality of the stream. Some species thrive in conditions of low oxygen and high organic matter, whereas others must have water with high oxygen concentrations and low levels of organic material. Collect samples of the bottom, sort through the material, and collect organisms for identification in the lab. Also collect some of the sediment from the bottom to return to the lab, where organisms can be sorted under a microscope. Identify organisms using the keys provided by your instructor.

    B.   Standard Plate Count (Group VI)

A standard plate count is a method of assessing the total number of several kinds of bacteria in a water sample. Not all bacteria will grow under the conditions used, but most kinds will and a standard plate count gives an index of bacterial numbers.

        1.   Use a sterile bottle to collect a sample of water from the stream. Collect away from the bank and avoid collecting sediment from the bottom. The samples should be processed before six hours have elapsed. If a longer transit time is anticipated, the sample should be refrigerated at $10°$ C or less and in any case should be processed before thirty hours have elapsed. The following procedures should not take place in direct sunlight.

        2.   Obtain a sterile water blank containing 99 ml of sterile water.

        3.   Use a sterile pipette to transfer 1 ml of the water sample to the 99 ml sterile water blank. Mix the sample with the water in the water blank.

        4.   Obtain four sterile petri dishes labeled 1 ml, 0.1 ml, 0.01 ml, and 0.001 ml.

        5.   Use a sterile pipette to transfer 1 ml of the original water sample to the petri dish labeled 1 ml. Use the following technique to do so. Lift the lid of the petri dish just enough to allow the pipette to deliver the water to the sterile empty dish.

        6.   Use the same technique to add water to the other petri dishes as follows.

            a.   0.1 ml from original sample to dish labeled 0.1 ml

            b.   1.0 ml from mixed diluted sample to dish labeled 0.01 ml

            c.   0.1 ml from mixed diluted sample to dish labeled 0.001 ml

        7.   Melt the tryptone glucose extract agar and hold at $44-46°$ C until used.

        8.   Lift the lid of each petri dish in turn just high enough to allow you to pour 10–12 ml of the agar into the petri dish. Swirl the contents gently to mix the water sample with the agar.

        9.   You should also make some petri plates using the sterile water and the medium to see that they were not contaminated.

      10.   Incubate the petri dishes for forty-eight hours at $32°$ C.

      11.   After 48 hours count the number of bacterial colonies growing on each plate. Since you know the size of the original sample, you should be able to determine the number of bacteria per milliliter of the original water sample. (Ideally, one of your plates should have between 30 and 300 colonies. Use this plate for determining the number of bacteria per ml.)

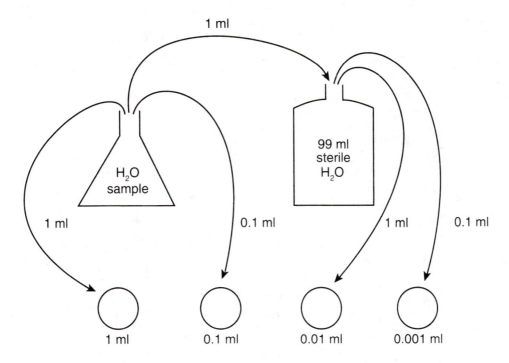

C.   Coliform Bacteria (Group VII)

Coliform bacteria are found in the intestines of humans and other animals; therefore, the presence of these kinds of bacteria is an indication of contamination from human or animal waste products. The coliform bacteria themselves are not a hazard but indicate that other pathogenic (disease-causing) bacteria may also be present. If the source of the coliforms is human, then we can assume that some human pathogens will be present.

1.   Use a sterile bottle to collect a sample of water from the stream away from the bank. Be careful not to collect sediment from the bottom. Stopper the bottle. If the sample cannot be used immediately (within one hour) it should be cooled to 10° C or less until it can be used. In any case, it should be used within six hours.

2.   Obtain nine fermentation tubes containing lactose broth. Three should be labeled 10 ml, three should be labeled 1 ml, and three should be labeled 0.1 ml.

3.   Shake the water sample thoroughly.

4.   Use a sterile pipette to place 10 ml of the water sample into each of the tubes labeled 10 ml, 1 ml into the 1 ml tubes, and 0.1 ml into the 0.1 ml tubes.

5.   Incubate at 35° C for 24 hours.

6.   After 24 hours examine the tubes for the presence of gas in the small inverted vial.

7.   Examine again at the end of 48 hours.

8.   Record all the tubes that show a positive test (i.e., gas has collected in the vial *and* the culture is cloudy).

9.   Consult a most probable number table to determine the approximate number of coliform bacteria in a 100 ml sample. (Your instructor will provide a most probable number table.)

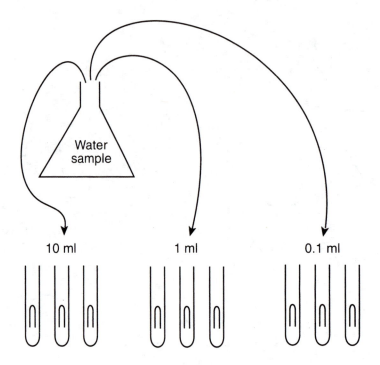

Stream Ecology
Data Sheet

Name _____

Section _____

	Stream 1	Stream 2
Temperature (°C)		
pH		
Suspended matter (mg/l)		
Oxygen concentration (ppm)		

Phosphate
    Orthophosphate (mg/l) ____
    Metaphosphate (mg/l) ____
    Organic phosphate (mg/l)		

Nitrogen
    Ammonia (mg/l) ____
    Nitrate (mg/l) ____
    Nitrite (mg/l)		

Organisms identified		
Standard plate count (number/ml)		
Coliform bacteria (number/100 ml)		

1.   List three water-quality differences you think are typical of a polluted stream and an unpolluted stream.

2.   Why is it important to determine the number of coliform bacteria in a water source?

3.   What kinds of organisms were typical of the polluted stream? Were they different from those found in the unpolluted stream? What does this indicate?

Stream Quality Assessment

Objectives

1. Understand some of the more fundamental measures of stream quality and water quality.
2. Understand some of the factors that can lead to reduced water quality.
3. Investigate the effects of reduced water quality on aquatic life.

Safety Considerations: Since sampling will take place in a stream, use standard water safety procedures.

Introduction

Water is a key component to life throughout the world. Indeed, it is clear that many of the richest and most diverse land-based ecosystems are those that receive a substantial amount of rainfall annually. In aquatic ecosystems, water availability and water quality are even more essential to the health and viability of the biological systems. Water resources are so important to human activity and quality of life that it has been predicted that the next wars will be fought over water resources and not oil.

Procedures

Equipment you will need:

   a.   Needle-nose forceps
   b.   White collecting pan
   c.   Several jars of different sizes
   d.   An old window screen with no holes approximately 1 meter by 1 meter
   e.   Old tennis shoes for wading in streams
   f.   A dissecting scope or magnifying glass

1. Position the screen vertically at right angles to the flow of water so that the water flows through the screen. If the stream has a firm, sandy bottom, place the screen so that it is downstream from an area of the stream bottom that is covered in part with sand, and in part with leaves, sticks, or other debris. Be certain that the bottom edge of the screen is flush with the bottom of the stream, allowing no insects to escape. Also, do not allow any water to flow over the top of the screen, as this could also allow insects to escape.
2. Measure a distance one meter upstream from the screen. In this area zero to one meter upstream from the screen, pick up the rocks, brush them with your hands and let the loose organisms, debris, and other matter float down into the screen. If the bottom is sandy or muddy, agitate this material lightly by kicking the streambed in a diagonal motion toward the screen for 30 seconds. After kicking the streambed, wait 2 minutes while the water flow carries the debris into the screen.
3. Once you have waited the 2 minutes to allow material to collect in your screen, pull the screen out of the stream so as to keep the material collected on the screen. Put the screen on a light-colored, flat surface and pull all the material from the screen with your forceps and place it in the white collecting pan with sufficient water to allow the animals to survive. Look very closely at the material in the collecting pan and record any objects that are moving animals. Most of the organisms will be insects.
4. Once you are certain that you have removed all the insects from the screen, count the total number you have collected. Separate the insects into groups that look generally similar in terms of body style, number of legs, wings, or tail shape. Place the separated groups into collecting jars that are also partially filled with stream water.
5. To evaluate the quality of the stream, we will tabulate the numbers of insects you found in the one-square-meter section of the streambed. To tabulate this information, answer the following questions:

   a.   What are the total number of insects found in the one-square-meter area?
   b.   Other than the heads, how many organisms are white, red, or gray?
   c.   Other than the heads, how many organisms are black, green, brown, or tan?
   d.   How many different species of organisms can you identify?
   e.   Is one organism most common, and if so, how many can you count?
   f.   Take a walk along 30 meters of the stream and note if you see any fish.

Stream Quality Criteria Based on Insect Survey

Most often, sophisticated analytical chemistry is used to evaluate the physical quality of a water resource such as a lake, river, stream, or groundwater. In this lab we will use the insect counts you have collected to make some general statements about the quality of the stream.

According to the number of insects per square meter and the number of fish visible, streams can be rated from good to dead. Please note the following definitions:

Toxicity: Chlorine, acids, heavy metals, pesticides, and other pollutants increase the toxicity of the stream and reduce the numbers of insects and fish in the water. A toxic problem is typically the only reason a water system will be totally devoid of insects.

Physical: Physical conditions such as heat and sedimentation can reduce the quality of the stream for insects and fish. Increased amounts of sediments in the water can lead to less visibility, and increased silting in the streambed. This reduces feeding opportunities for organisms that use vision to locate food. Silting can cover and destroy valuable rocky and pebbled stream-bed habitats.

Organic: Organic pollution such as livestock wastes, agricultural fertilizers, and human waste flows can have drastic effects on stream ecology. Organic pollution usually results in increased oxygen demands thereby changing the stream so that it becomes dominated by life forms such as trash fish, worms, and bacteria, that can survive low oxygen concentration.

Stream Quality Categories

Dead:
A.   Less than one organism found per square meter sampled.
B.   No fish observed.

Note: Clear water might indicate a toxicity problem.

Poor:
A.   More than one insect found per square meter.
B.   90% of organisms found are gray-white or red in color.
C.   One to two types of organisms are distinguishable.
D.   No fish observed.

Note: Speculate that there might be moderate levels of toxic substances present or a severe physical problem.

Fair:
A.   More than one insect found per square meter.
B.   11–30% of organisms found are black, green, brown, or tan in color.
C.   Three to six types of organisms are distinguishable.
D.   There is not an organism that makes up more than 90% of the sample.
E.   Fish are observed.

Note: Speculate mild toxicity, moderate physical problems, or excessive organic matter present.

Good:
A.   More than one insect found per square meter.
B.   At least 30% of organisms found are black, green, brown, or tan in color.
C.   Six or more types of organisms are distinguishable.
D.   There is not an organism that makes up more than 50% of the sample.
E.   Fish are observed.

Note: No problems appear noticeable.

Stream Quality Assessment
Data Sheet

Name _____

Section _____

1. What are some of the trends you see in the river characteristics for the different quality ratings of dead, poor, fair, and good?

2. Were you surprised at the number of organisms you found?

3. What overall effect do you think a poor to dead stream has on its surrounding wildlife? What effect do you think poor stream quality may have on humans?

4. What are some common management techniques to improving stream quality?

Ecological Stress and Species Diversity

Objectives

1. Understand the relationship between ecological stress and species diversity.
2. Understand why aquatic macroinvertebrates may be used as indicators of water quality.
3. Identify the taxa of macroinvertebrates from several sites of differing water quality.

Safety Considerations: Given that the actual sampling will be along watercourses, care should be given to basic water safety.

Introduction

The benthos consists of the organisms that inhabit the bottom substrate of lakes, ponds, and streams. Benthic organisms play several important roles in the aquatic community: they are involved in the important process of mineralization and recycling of organic matter, and they are vital links in the trophic sequence of aquatic communities. Many benthic insect larval forms are major food sources for fish.

Quantitative benthic studies usually determine the number and kinds of organisms present and often serve as indicators of water quality. Benthic organisms are especially useful in pollution studies, and indices of their diversity are widely used for estimates of the degree of pollution—low species diversity indicates water of low quality.

This exercise focuses on the use of benthic macroinvertebrates as indicator organisms for the following reasons:

1. The importance of the bottom fauna to commercial and sport fishing.
2. Their well-known sensitivity and quick response to environmental stress and degradation.
3. The complex and long life cycle (about one year) enables the effects of stress over time to be studied.
4. Their sessile nature lends itself well to use in a small sampling area, where they can serve as natural monitors of water quality.

Procedures

1. Students will collect benthic macroinvertebrates from several locations using artificial substrates.
2. Students will identify the specimens collected and examine the areas sampled for species diversity.
3. Discuss the relationship between ecological stress and your analysis of the aquatic macroinvertebrate communities studied.
4. Prepare and hand in the data sheet at the end of the exercise.

Collection Methods

Many types of artificial substrates, including bricks, building tiles, and concrete blocks have been used for aquatic macroinvertebrate studies. These materials have the advantage of having regular, known surface areas that can be cleaned for quantitative study. Whichever substrate you choose to use, make sure it is of relative uniformity and is free of paint or other surface contaminants that would inhibit colonization.

Site Selection

The number of sites chosen for analysis will vary. If possible, choose sites that are likely to differ in water quality, such as upstream and downstream from a factory. You could also choose different aquatic environments for study, such as a large lake, a small lake, and a river. You may want to study an urban river and a rural river system for comparison.

Sampling

1. Number your artificial substrates with permanent ink.
2. Place the artificial substrates in selected sites. Secure substrates with anchors, ropes, or poles, and mark the areas with floats if necessary.
3. Allow artificial substrates to colonize for a minimum of two weeks. Colonization time may need to be longer depending on the season of the year, nature of the study, and water temperature.
4. After colonization gently remove the artificial substrate. Place the substrate onto a U.S. Standard #30 sieve so that the organisms will not be lost. Scrape the substrate, then sieve to collect the organisms. Place the organisms in a large-mouthed jar to be returned to the laboratory for live examination if possible, or preserve in alcohol if necessary.

Analysis

1. In the laboratory or classroom, select a known portion of each sample for examination.
2. Use a macroinvertebrate key and a dissecting microscope to identify the taxonomic family to which each kind of organism belongs. The drawings in the figure should help you identify groups.
3. Count the number of each kind.
4. Count the number of families.
5. Record on data sheet.

Sequential Comparison Index

A sequential comparison index (SCI) can be used to assess species diversity. This method is particularly valuable because it is not necessary to identify organisms. All that is required is that you be able to tell that one organism is different from another.

To determine a sequential comparison index, randomly choose individuals from the collection and determine whether it is the same as or different from the one preceding it. Give each kind of organism a symbol. Each time the succeeding organism is different from its predecessor, begin a new run. Record the number of runs and the number of individuals in your sample. For example, if there were five kinds of organism A, B, C, D, and E and they were selected in the following order,

$$\underset{\text{run 1}}{\underline{D\ D}} \quad \underset{\text{run 2}}{\underline{E\ E\ E}} \quad \underset{\text{run 3}}{\underline{A}} \quad \underset{\text{run 4}}{\underline{D\ D}} \quad \underset{\text{run 5}}{\underline{B\ B}} \quad \underset{\text{run 6}}{\underline{C\ C\ C\ C\ C}}$$

there would be six runs and fifteen individuals.

$$\text{The sequential comparison index} = \frac{\text{number of runs}}{\text{number of individuals}} = \frac{6}{15} = 0.4$$

The closer the number is to 1, the more species diversity exists. If you had fifteen organisms in your sample and each was of a different species, you would have fifteen runs and fifteen individuals and the sequential comparison index would be 1.

1. Combine all the organisms collected from one site. Arrange the animals randomly in a straight line in an enamel pan and determine the sequential comparison index.
2. Record the diversity index for each sample.

Macroinvertebrates Identification Chart

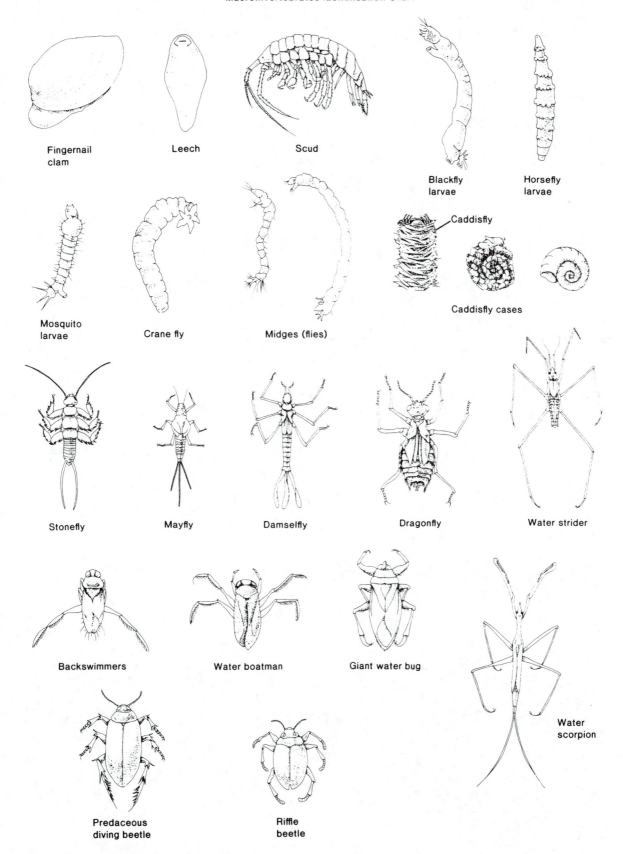

Only the most common organisms are shown as immatures. The actual organisms may be smaller or larger than shown, due to species differences and maturation stage at collection time.

Ecological Stress and Species Diversity Data Sheet

Name _____

Section _____

Number of Organisms

	Station 1	Station 2	Station 3	Station 4	Station 5	
Family						Common name
Total individual organisms						
Number of families						
Diversity index						

1. In an essay format, discuss the information gained from analyzing the substrate samples. From your experience, could you link species diversity to water quality?

2. As a class exercise, evaluate the diversity of your community using such criteria as ethnic, racial, religious, and socioeconomic groups; life-styles; industries; and landscape features and forms. What elements of diversity have proven troublesome? What additional elements of diversity would improve your community?

Behavioral Differences in Small Mammals

Objectives

1. Construct a hypothesis relating to the behavior of small mammals toward objects in their environment.
2. Test the hypothesis by using an arena that has walls as physical objects.

Safety Considerations: The animals may bite if they are not handled gently. Choose a person who is confident in handling animals to place them onto the arena. He or she may want to wear gloves.

Introduction

Most animals make use of objects in their environment. Birds use twigs and grass to build nests, small mammals build burrows in grass and earth, birds of prey use trees and electrical lines as observation places, and many insects hide under leaves and bark. The use of physical objects for cover by small mammals is common but must be coupled with the typical life history of the animal. Diurnal (active during the day) animals have a greater chance of being observed by predators than do nocturnal (active at night) animals. Small forest or prairie mammals can conceal themselves more easily as they go about their normal activities than can desert animals that live in areas with little grass or other vegetation to conceal them. Does this mean that small desert mammals would behave differently from other small mammals?

The common laboratory mouse has been used for years as a research animal. It has been bred to do well in laboratory conditions, to show little genetic variability, and to be handled easily by people. Does this mean that laboratory mice have behavior patterns that differ from those of other mice of the same species? Would laboratory mice respond to physical objects in their environment differently from wild or free-living mice?

Preview

During this exercise you will

1. Form a hypothesis relating to the behavior of small mammals toward walls in their environment
2. Use an arena with walls to test the behavior of one or more species of small mammals
3. Use the chi-square test to determine if the results of the test are significant

Procedure

Your instructor will probably divide you into groups of about three persons. One person will need to be a timekeeper, one person will need to be a recordkeeper, and another person will need to be the animal handler.

Your instructor will give you two kinds of small mammals. They may be different strains of the same species or they may be of different species. Your instructor will also give you some background information on the animals: what kinds of ecological niches they occupy in nature, whether they are wild or laboratory animals, and so forth.

From the information you are given, construct a hypothesis concerning how these two kinds of small mammals will differ in the amount of time they will spend adjacent to objects.

Write your hypothesis in the space provided on the data sheet.

To test your hypothesis we will use an arena that consists of a flat sheet of material with thirty-six squares drawn on it. The four central squares will be occupied by boxes or will be walled off from the two rows of squares surrounding the central squares. A diagram of this arena is shown in figure 15.1. There will be no walls at the periphery of the arena. Therefore, the arena should be elevated so that the small mammals will not wander off the arena. If one of them does jump off, gently capture the animal and return it to the same square from which it left the arena.

Since you want to study the behavior of the animals toward objects in their environment, you want to prevent other kinds of stimuli from influencing their behavior. Dim the lights in the laboratory and refrain from loud noises or sudden movements during the experiment.

Place an animal on the arena in one of the peripheral squares. Allow it to wander about the arena for ten minutes. Every fifteen seconds, mark the spot where the animal is situated at that point in time. The timekeeper should quietly say "now" every fifteen seconds so that the recorder knows when to make the check mark. Repeat with as many animals as you can, but be sure that you test equal numbers of the two kinds of animals. Record the data in a way that will allow you to test your hypothesis. You may only need to know whether the test animal was in the central area or in peripheral areas, or you may need more detailed information about location.

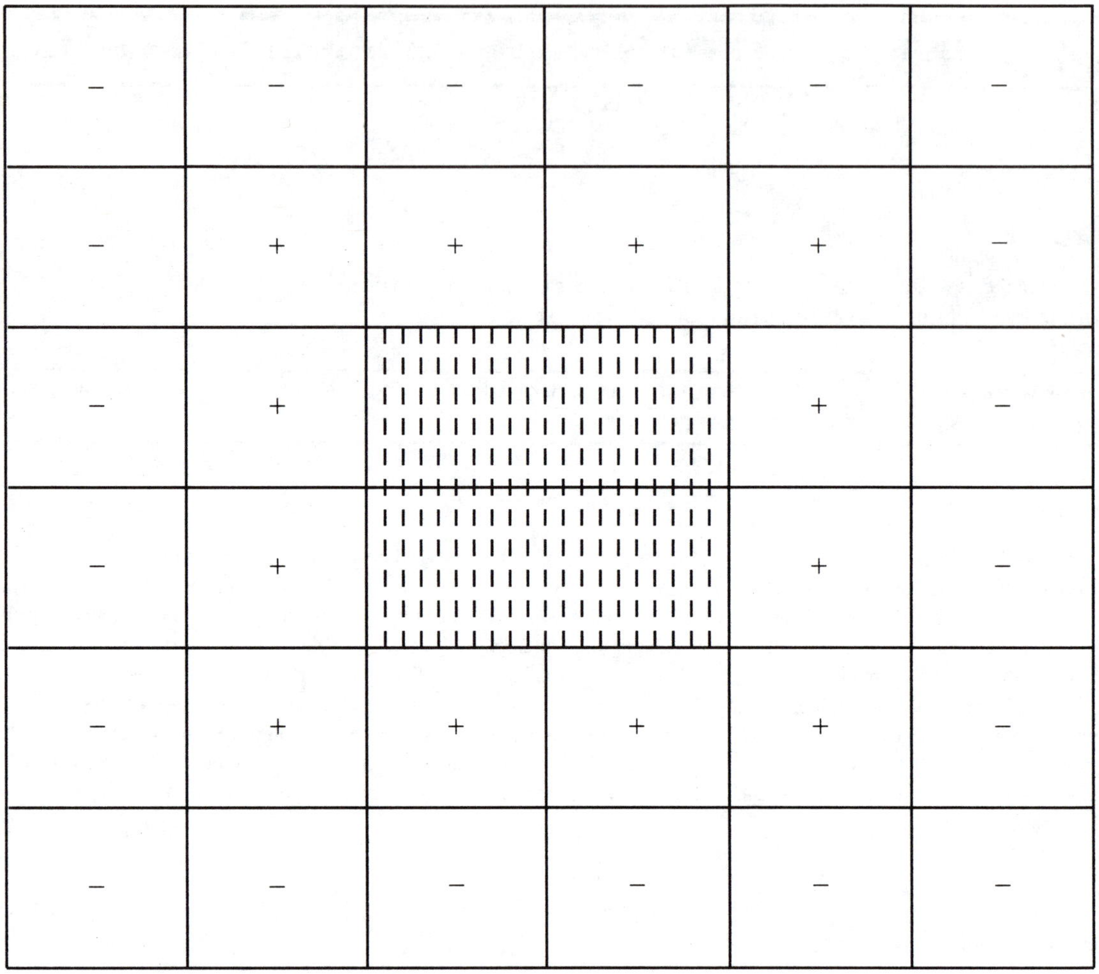

Figure 15.1 Diagram of arena to use to test your hypothesis.

Behavioral Differences in Small
Mammals Data Sheet

Name _____

Section _____

Hypothesis:

Analysis of Results

You will want to pool all of the data collected by the members of your class before beginning the statistical analysis of your results.

There are twenty (20) peripheral squares and only twelve (12) central squares; therefore, if the animals were to move about the arena at random you should expect that they would spend 20/32 (62.5%) of their time in the peripheral squares and 12/32 (37.5%) of their time in the central squares next to the barrier.

You can determine if the animals moved about the arena at random by comparing the expected against the observed. To help determine if the results are significantly different from random, use the chi-square test found in Appendix 4.

To determine the expected in the periphery if the animals moved at random, count the total number of squares checked. Multiply this number by 62.5%.

Total number of squares checked × 0.625 = _____ if the animals moved about the arena randomly.

To determine the expected in the central squares, multiply the total squares checked by 37.5%.

Total number of squares checked × 0.375 = _____ if the animals moved about the arena randomly.

1. Did either of your kinds of animals prefer the central squares?

   You can also use the chi-square test to determine if the two kinds of animals behaved significantly different. Choose one of the kinds of animals __(A)__ and make it the "expected" category. Calculate the percentage of time this kind of animal __(A)__ spent in the peripheral squares.

$$\frac{\text{Total peripheral squares checked}}{\text{Total squares checked}} = \% \text{ time } \underline{\text{(A)}} \text{ spent at periphery}$$

   Multiply this percentage by the total squares checked for animal __(B)__. This is the expected for __(B)__ if __(B)__ were the same as __(A)__.
   You can calculate the "expected" for the central squares in a similar manner.

2. Did your two kinds of animals differ significantly in the amount of time they spent in the central squares?

3. Was this expected or not?

4. Do you need to revise your hypothesis?

Objectives

1. Measure species diversity in a natural and human-dominated ecosystem.
2. Assess the role of humans in altering species diversity.

Species Diversity

Natural ecosystems have a great deal of variety. Many different species interact with one another to create a stable, persistent, functional unit called an **ecosystem.** Disturbed ecosystems tend to have large fluctuations in population size and often have a reduced number of kinds of organisms, although some species may have very large numbers. Species diversity is, therefore, a convenient method of assessing the health of an ecosystem. In this exercise we will compare a relatively undisturbed ecosystem, such as a prairie, woodlot, or unpolluted stream, with a similar but intensely managed ecosystem such as a lawn, agricultural crop, forest plantation, irrigation canal, or drainage ditch.

Procedure

1. Choose two ecosystems to assess. These can be suited to your local situation. Just make sure they are similar but that one is under intense human influence. Your instructor will probably have suggestions.
2. A sequential comparison index will be used to assess species diversity. This method is particularly valuable because it is not necessary to identify organisms. All that is required is that you be able to tell that one organism is different from another.
3. Select a method for collecting organisms.
    —   Soil and litter can be brought into the lab and the organisms can be collected in a white enamel pan. (The tiny organisms show up better against a white background.)
    —   A Berlese funnel can also be used to collect soil organisms. (Your instructor can assist you in setting up the Berlese funnel.)
    —   Pit traps consisting of tin cans or plastic containers can be set in the ground even with the surface to capture small organisms such as ground beetles, snails, millipedes, and centipedes.
    —   Sweep nets can be used to capture insects in grassy areas.
    —   Dip nets can be used to assess aquatic habitats.
    —   Birds can be identified by direct observation. Simply watch a given area and log in sequence each bird seen.
    —   Plants can be collected from randomly chosen plots.
    —   Other methods can be used depending on the ecosystems chosen to assess and the equipment available. Your instructor may wish to have several methods used.
4. Return to the lab with your organisms. Combine all the organisms collected from one site. Remove individuals randomly from your collection and calculate a sequential comparison index. (In some cases you may be able to do this process in the field and the organisms can be released. In other cases you may need to inactivate the organisms by chilling or killing in alcohol.)
5. To determine a sequential comparison index, randomly choose individuals from the collection and determine whether it is the same as or different from the one preceding it. Give each kind of organism a symbol. Each time the succeeding organism is different from its predecessor, begin a new run. Record the number of runs and the number of individuals in your sample. For example, if there were five kinds of organism A, B, C, D, and E and they were selected in the following order,

$\underline{D\ D}$	$\underline{E\ E\ E}$	$\underline{A}$	$\underline{D\ D}$	$\underline{B\ B}$	$\underline{C\ C\ C\ C\ C}$
run 1	run 2	run 3	run 4	run 5	run 6

there would be six runs and fifteen individuals.

$$\text{The sequential comparison index} = \frac{\text{number of runs}}{\text{number of individuals}} = \frac{6}{15} = 0.4$$

The closer the number is to 1, the more species diversity exists. If you had fifteen organisms in your sample and each was of a different species, you would have fifteen runs and fifteen individuals and the sequential comparison index would be 1.

Human Influence and Species Diversity Data Sheet

Name _____

Section _____

Kind of ecosystem	Number of runs	Number of individuals	Sequential comparison index

1. Which ecosystem had the greatest species diversity?

2. Are your results similar to those of other groups of students in the class?

3. Describe three ways that human activity could have influenced species diversity.

Human Population—Changes in Survival

Objectives

1. Understand differences in human mortality and survivorship between historic and modern times.
2. Understand how changes in human mortality and survivorship have influenced population growth.

Safety Considerations: There are no specific safety considerations in this exercise.

Introduction

The survival rate of humans in North America has increased significantly in the past 100 years. Improved nutrition, preventive medicine, life-style changes, and new technologies are a few of the reasons for this improved life expectancy. Increasing life expectancies have had an impact on population growth rates. Put simply, there are more of us and we are living longer. In Rome during the first to fourth centuries the expectation of life was about 22 years at birth. Today the expectation is approximately 75 years at birth in North America. Of particular note is the decline in infant and youth mortality in North America during the past 100 years.

Procedures

1. Record the age at death for males and females in your community who died before the year 1900.
2. Using the obituary page from your local newspaper, record the age at death for males and females in your community who died during the past five years.
3. Enter the data on the data sheet.
4. Plot the data on a graph (survival curve) for both males and females for the two time periods.
5. Analyze the data and the reasons for change.

Directions

You will obtain two sets of data giving numbers of deaths in your community by age. The first set, representing vital statistics of your region in pre-1900 historic times, will be obtained from one or more graveyards. The second set, representing current mortality figures, will be obtained from the obituary section of your local newspaper.

In order to prevent duplication of data, discuss in class what cemeteries are to be included in your study. Try to prevent the same cemetery (or section of a cemetery) from being counted more than once. Each student should record as many gravestones as necessary to give a total class sample of at least 100 males and 100 females. If you wish to do this exercise on an individual basis you need not have such a large sample. Using the pre-1900 data sheet, record your entries for both males and females.

Use the data sheet and obtain information on the deaths of 100 males and 100 females representing current deaths from the obituary page of your local newspaper over the past five years.

To determine a survival curve use the following method. Since you have data on 100 pre-1900 males, 100 pre-1900 females, 100 current males, and 100 current females, you can determine the number surviving to each age for each of the four groups by using the following technique.

EXAMPLE

Age at death (years)	Number dying	Number surviving			
0	0	100 –0	=	100	Plot the underlined numbers on the graph.
0–0.99	10	100 – 10	=	90	
1–4.99	15	90 – 15	=	75	
5–9.99	12	75 – 12	=	63	

Use these data to graph a survival curve for each of the four groups:

pre-1900 females
pre-1900 males
current females
current males

Use different-colored pens or pencils to record each of the sets of data on the graph on the data sheet.

Human Population—Changes in Survival Data Sheet

Name _____

Section _____

Cemetery

Age at Death (years)	Male	Female
0 – 0.99		
1 – 4.99		
5 – 9.99		
10 – 14.99		
15 – 19.99		
20 – 24.99		
25 – 29.99		
30 – 34.99		
35 – 39.99		
40 – 44.99		
45 – 49.99		
50 – 54.99		
55 – 59.99		
60 – 64.99		
65 – 69.99		
70 – 74.99		
75 – 79.99		
80 – 84.99		
85 – 89.99		
90 – 94.99		
95 – 99.99		
100+		

Obituary

Age at Death (years)	Male	Female
0 – 0.99		
1 – 4.99		
5 – 9.99		
10 – 14.99		
15 – 19.99		
20 – 24.99		
25 – 29.99		
30 – 34.99		
35 – 39.99		
40 – 44.99		
45 – 49.99		
50 – 54.99		
55 – 59.99		
60 – 64.99		
65 – 69.99		
70 – 74.99		
75 – 79.99		
80 – 84.99		
85 – 89.99		
90 – 94.99		
95 – 99.99		
100+		

1.   What accounts for the difference in the four curves?

2.   What do you think the curves would look like in the next century? What factors could influence the curves?

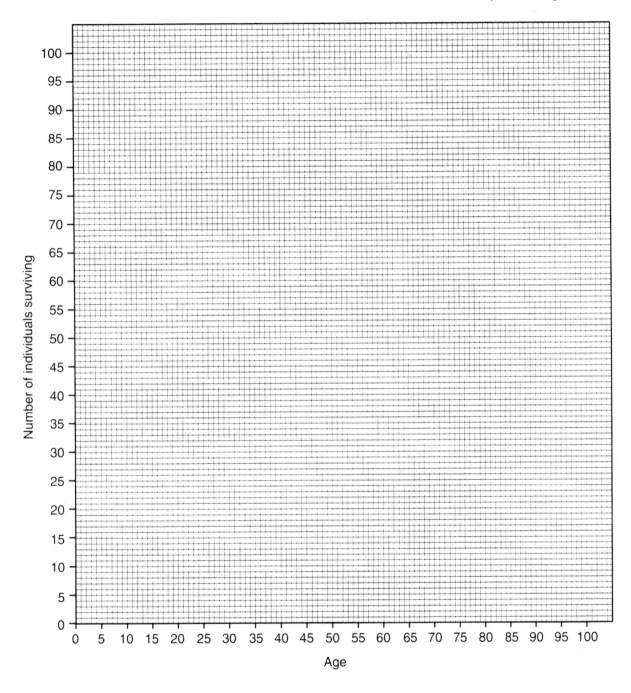

Population Demographics

Objectives

1. Become familiar with the standard criteria used for characterizing populations including age distribution, birthrates, death rates, and male/female ratios.
2. Construct population age structures for assorted countries and draw conclusions about population growth potential.

Introduction

The impact humans have on the earth is often characterized by the following equation:

$$I = P \times C \times T$$

I = Impact on natural earth systems per unit time
P = Human population
C = Consumption per person per unit time
T = Technology factor (high for environmentally destructive technology, low for environmentally friendly technology)

It makes sense that larger numbers of people will have a greater effect on earth systems than fewer people. The equation ($I = P \times C \times T$) captures this relation in a simple expression. Any change in the population (P) will have a major effect on the impact (I) unless consumption (C) or technology (T) change significantly.

The study of population demographics concentrates on the descriptive characteristics of human populations including issues such as changes in the size and structure of populations, and rates of population change. The worldwide human population is currently experiencing exponential growth, a condition in which the population increases at a faster and faster rate. As human population growth continues exponentially, the equation ($I = P \times C \times T$) would indicate that the resulting impact on earth's natural systems would also increase exponentially (unless consumption per person is reduced or environmentally friendly technology replaces current technology).

Demographers use the following equations to calculate the key factors that describe changes in population characteristics:

$$\text{crude birthrate (births /1,000)} = \frac{\text{live births per year}}{\text{mid-year population}} \times 1,000$$

$$\text{crude death rate (deaths /1,000)} = \frac{\text{deaths per year}}{\text{mid-year population}} \times 1,000$$

$$\text{annual rate of population change \%} = \frac{\text{birthrate} - \text{death rate}}{10}$$

$$\text{population doubling time (years)} = \frac{70}{\text{annual rate of population change (\%)}}$$

The population growth rate has been correlated to a number of factors which affect the birthrate, the death rate, or both. The following factors appear to influence the birthrate:

a. Level of education and wealth
b. Importance of children for family labor purposes
c. Urbanization—higher birthrates in rural areas
d. Cost of raising children
e. Education and employment opportunities
f. Average age at marriage
g. Availability of birth control devices
h. Cultural norms

Factors that influence death rates:

a. Nutrition
b. Sanitation
c. Advances in available health care
d. Ability to afford medical care

Population Demographics
Data Sheet

Name _____

Section _____

1. Use the table below and the equations presented on page 93 to calculate the annual rate of population change and the population doubling time for each of the fifteen countries.

Country	Population density (people/ square mile)	Crude birthrate (births/ 1,000)	Crude death rate (deaths/ 1,000)	Annual rate of population change (%)	Doubling time (years)
1 Afghanistan	69	49	22		
2 Bangladesh	2,266	37	13		
3 Dominican Republic	408	23	6		
4 France	272	13	9		
5 Guatemala	240	39	7		
6 Hungary	289	12	14		
7 Mexico	122	29	6		
8 Netherlands	1,164	13	9		
9 Norway	36	14	11		
10 Russia	23	12	11		
11 Pakistan	411	44	13		
12 Spain	203	10	9		
13 Tanzania	81	46	15		
14 United States	73	16	9		
15 Zimbabwe	72	41	11		

(Data from 1993 World Population Data Sheet, Population Reference Bureau, Inc)

2. Use the population density information given and the annual rate of population change you calculated to plot the relationship on the graph provided.
    Does the annual rate of population change appear to be related to population density? (Hint: The two factors, population density and annual rate of population change, are very strongly correlated if you can draw one straight line and connect all the points. If you can draw a line that most of the points cluster near the two factors are correlated less strongly. If there is no straight line that the points cluster near the two factors are not correlated.)

3. Draw a graph of the relationship between the doubling time and the crude birthrate on the graph provided. On the same graph plot the relationship between crude death rate and doubling time. Plot the two sets of data with different colors. Which of the two (birthrate or death rate) is most closely related to doubling time (i.e., which of the two sets of data is closer to a straight line)?

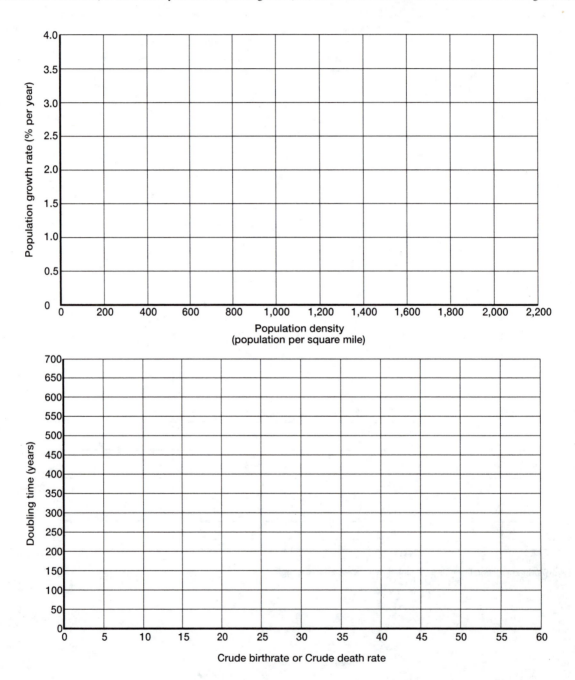

Estimating Population Size

Objectives

1. Use a technique for estimating populations.
2. Understand the factors that can influence the accuracy of a population estimate.

Safety Considerations: Your instructor will give specific instructions regarding hazards.

Introduction

Scientists often estimate the size of a population of animals by sampling the population. If a specific number of individuals are captured, marked, and released into the wild population, it is possible to estimate the total population by using the following ratio:

$$\frac{\text{total population}}{\text{number marked initially}} = \frac{\text{total in 2nd sample}}{\text{number of marked individuals in 2nd sample}}$$

This ratio can be rearranged as follows:

$$\text{total population} = \frac{\text{total in 2nd sample}}{\text{number marked in 2nd sample}} \times \text{number marked initially}$$

For example, if

1. 100 animals were captured, marked, and released;
2. mixed with the total population;
3. 50 animals are recaptured, of which 10 are marked.

$$\text{total population} = \frac{\text{total in 2nd sample (50)}}{\text{number marked in 2nd sample (10)}} \times \text{number marked initially (100)}$$

$$\text{total population} = \frac{50}{10} \times 100 = 500$$

Several assumptions must be met if this method is to be reasonably accurate:

1. Marked individuals must not behave differently from unmarked individuals.
2. Marked individuals must have time to mix freely.
3. The sampling of the population must capture animals randomly.

Procedures

1. Capture and mark organisms as instructed.
2. Return the organisms to their normal habitat and allow them to mix freely.
3. Resample the population.
4. Record total number in second sample and the number in the second sample that were marked.
5. Calculate population size.

Laboratory Exercise

Since it is possible to use this method with a variety of animals, your instructor will give specific instructions on how to proceed. Several possibilities exist.

1. "Tag" some of the students in the class as they enter. After the class has assembled, "sample" the class by counting every fifth student. Ask if they are "tagged" and calculate the size of the class. Since you know how many people are in the class you can check your accuracy.

2. Laboratory insect populations of meal worms, fruitflies, or flour beetles could be used. Sample the population and mark a known number of individuals with a colored pen. Return the marked individuals to the container, allow them to mix thoroughly, sample the population, and count the total number of individuals and the number of marked individuals. Calculate the size of the population.

3. Sampling wild populations of birds, mammals, or amphibians is also possible but usually requires special circumstances, equipment, and techniques. In many cases special permits are required from state or federal regulatory agencies. Your instructor will give specific instructions. Possibilities are

    (spring) breeding amphibians at a pond
    (summer/fall) live trapping of small mammals
    (winter) mistnetting birds at bird feeder

Estimating Population Size
Data Sheet

Name _____

Section _____

Number of marked individuals released _____

	Total captured	Number captured that were marked
Recapture sample		

The total number of animals in the population is _____.

List three factors that would make this method inaccurate.

Is Your Campus Friendly to Wildlife?

Objectives

1. Understand the basic needs of wildlife for food, water, and shelter.
2. Inventory your campus to determine its suitability for wildlife.
3. Develop plans to improve the wildlife habitat on your campus.

Safety Considerations: There are no specific safety considerations; however, use care when visiting areas not frequently traveled.

Procedures

1. Organize the class into work groups.
2. Survey the campus for sources of water, food, and cover for wildlife.
3. Develop a map showing where current resources are located.
4. Develop a plan for improving the wildlife habitat on your campus.

Initial Survey

Many kinds of wild animals do very well in urban settings as long as they are able to satisfy their basic needs for food, water, and cover to escape from predators and the elements. Survey your campus and make a map of the food and water sources and sheltered areas that could be used as cover. Look especially for the following kinds of situations:

1. Permanent water—ponds, swamps, bogs, lakes, creeks, irrigation ditches, etc.
2. Food sources—fruits and seeds of trees and shrubs, grass and weed seeds, bird feeders, wildflowers in bloom, garbage, greenhouse refuse, gardens
3. Cover—thickets of dense shrubs, open weed fields, hedgerows or fence lines, ditch banks, ravines, rock piles, landscape timbers, birdhouses, hollow trees, animal burrows, brush piles, landscape refuse, compost piles

Is Your Campus Friendly to Wildlife?
Data Sheet

Name _____

Section _____

Develop a plan to improve the quality of your campus for wildlife. Be particularly sensitive to providing ideas that do not require a high degree of maintenance or require substantial expenditure of money. For example, building and placing birdhouses around campus or allowing a field to grow into weeds would be preferable to suggesting that a pond be built or special food plots be planted every year. Obtain permission from the building and grounds department and implement your plan.

Elements of the plan

Water	Food sources	Cover

Field Trip Suggestions

1. Visit a power plant: nuclear, coal-fired, wood-fired, or hydroelectric. Determine the quantity of electrical energy produced, the size of the service area, if electricity is sold to other utilities, and whether the facility is used primarily for the base-load or peak-load requirements of the utility. Describe how the utility minimizes its negative ecological impacts.
2. Visit your school's power plant. Determine the amount of energy used per year, and calculate energy use per student per day. What steps has your school taken to reduce energy consumption? Why were they taken?
3. Visit a coal mine, oil field, or gas field. Describe how the company minimizes its negative ecological impacts.
4. Visit an energy information center. (Most power companies provide such services.) Describe five changes you would make and how much energy you would save. What would this mean in monetary terms?
5. Visit a pipeline or powerline right-of-way. Describe three ways the vegetation in the right-of-way differs from the adjacent, less disturbed land.
6. Visit an oil refinery. Describe how the company minimizes its negative ecological impacts. Determine where the company sells its product.
7. Visit a nuclear facility (hospital, X-ray installation, or nuclear power plant). List ten steps taken to assure safety.

Alternative Learning Activities

1. Invite a power company executive to talk to the class.
2. Invite the school's physical plant director to talk to the class about the energy requirements of the school and the costs involved.
3. Trace the path of oil from production facility to gasoline station by interviewing or writing letters to people and asking where they purchase their product (i.e., gas station gets its gas from a distributor, who gets it from a wholesaler, refiner, oil pipeline company, etc.).
4. Trace the path of coal to a power plant, uranium to a nuclear power plant, natural gas to a home.
5. Draw up a list of ten ways you could reduce energy expenditures. Implement your list.
6. Within your class conduct a contest to see who can reduce energy expenditures the most.
7. Collect all the wastepaper in a particular building or area of your campus and determine how much energy this represents in terms of calories of heat energy.
8. Visit a local hospital to learn how it generates low-level radioactive waste. How does it dispose of this waste? What does it cost?

The Effectiveness of Insulation

Objectives

1. Test the insulating effectiveness of various kinds of materials.
2. Compare results with published *R*-values for different materials.

Safety Considerations: The light bulb in the box will become hot. The apparatus should not be left unattended, or it may overheat. Check the electrical connections before turning on the bulb.

Introduction

One of the most economical ways to reduce heat loss or gain in buildings is by using appropriate insulation in construction. It is also possible to install insulation in already constructed buildings to reduce heat loss or gain. Most insulating materials are rated as to their insulating value. The standard unit is called an *R*-value. It is a material's ability to resist the flow of heat through it. The higher the *R*-value the better the insulating ability. The reciprocal of $R(1/R)$ is a measure of the amount of heat energy in British Thermal Units (BTUs) that would pass through a piece of material 1 square foot in area in 1 hour when the temperature is 1° Fahrenheit higher on one side of the insulation than on the other. A BTU is the amount of heat energy necessary to raise 1 pound of water 1° F and is equal to 252 calories. The following table lists typical *R*-values for several kinds of materials.

Materials	*R*-value
No insulation	0
Single-pane glass	0.9
Double-pane glass	1.85
Triple-pane glass	2.8
1-inch wood	1–1.5
1-inch Fiberglass batts	3.1–3.7
1-inch Styrofoam	5.5

Procedures

Use the apparatus provided to test the insulating ability of several materials in the following way.

1. Insert thermometer in the end of the apparatus.
2. Place insulating material in the middle of the apparatus (see diagram).
3. Close the lid.
4. Turn on the light bulb.
5. Record the change in temperature every five minutes for thirty minutes.
6. Graph the data.

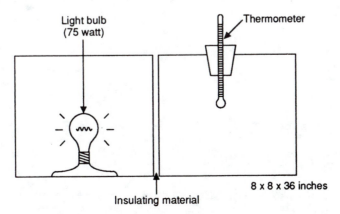

Light bulb (75 watt)
Thermometer
8 x 8 x 36 inches
Insulating material

Your instructor will probably divide the class into groups, with each group responsible for testing one or two different materials. In order to have a basis for comparison, at least one trial should be run with no insulating material in the apparatus. The other materials may be assigned by the instructor. Record all the data on the graph on the data sheet.

The Effectiveness of Insulation
Data Sheet

Name _____

Section _____

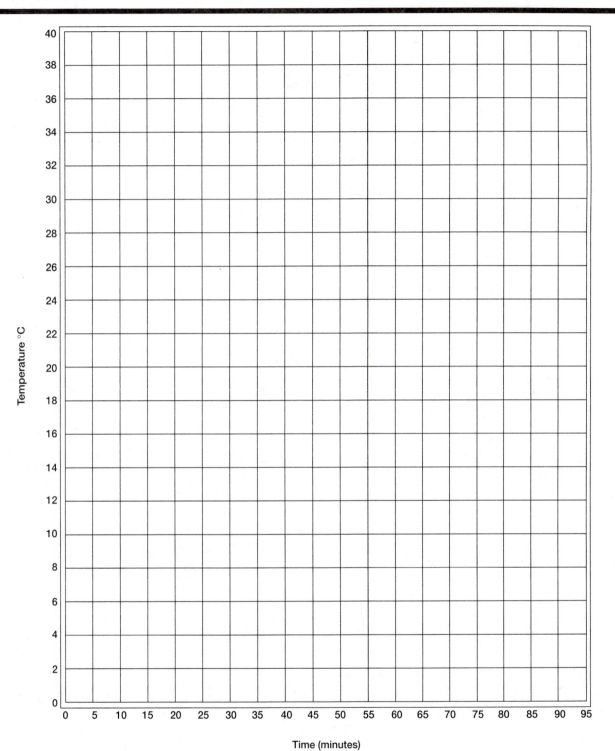

Temperature °C

Time (minutes)

1.   Which of the materials was the most effective insulation?

2.   If you needed to choose an insulation that was thin, which would you choose?

3.   Are your results consistent with published R-values?

Objectives

1. Calculate energy savings from changing window size or structure.
2. Calculate energy loss from a dripping hot water faucet.
3. Examine the implications of our life-style on energy consumption.

Safety Considerations: Care should be taken to unplug electrical appliances when moving or examining them.

Introduction

Except for the small countries of Luxembourg, Bahrain, Qatar, and Oman, North America uses more energy per person than all other parts of the world. This is true because historically we have always had abundant energy in the form of wood, coal, and oil. Because we have had a large supply, there has been less interest in developing ways to use energy more efficiently.

There are several categories of personal energy consumption that we all have some control over: heating and air conditioning, heating water, lighting, transportation, and purchase and use of electrical appliances.

The measurements used in North America to measure quantities of energy are quite diverse. Many kinds of heat measures are commonly given in British Thermal Units (BTUs), whereas electrical energy is usually measured in kilowatt-hours. Rather than try to convert all of the different units to the metric equivalent, we will use the standard units used in ordinary commerce.

BTUs in various amounts of fuel
1 gallon of fuel oil: 145,000
1 cubic foot of gas: 1,031
1 kilowatt-hour of electricity: 3,412
1 ton of coal: 25,000,000
1 cord of wood: 20,000,000
1 gallon of gasoline: 125,000

Procedure

In this exercise we will look at a variety of energy uses or misuses and calculate the energy that would be saved by a variety of changes in the way we do things.

Heating and Air Conditioning

One of the major ways that energy leaves or enters buildings is through windows. A single-pane window has an R-value of 0.9. $1/R$ is equal to the number of BTUs that would pass through a 1-square-foot surface in 1 hour if the difference in temperature on opposite sides of the surface is $1°$ Fahrenheit. Therefore we can calculate heat loss or gain through a window by using the following formula:

$$\text{heat loss/gain in BTU per hour} = \frac{\text{ft}^2 \times \text{difference in temp. }(°\text{ F})}{R\text{-value}}$$

Choose a single-pane window in your classroom or home and measure its surface area. Measure the difference in temperature between the inside and the outside of the window. Calculate the rate of heat transfer through the window by using the formula.

What effect would a $5°$ F decrease in the temperature on the inside of the window have on the rate of heat transfer?

What would happen to the rate of heat transfer if the size of the window were reduced by 50%?

Double-pane windows have an R-value of 1.85. Triple-pane windows have an R-value of 2.8. What would be the effect on the rate of heat transfer if the single-pane window were replaced with double- or triple-pane windows?

Obtain R-value for special low emission glass and calculate heat loss. Record results on data sheet.

Heating Water

Water resists changes in temperatures. In other words, it takes a lot of heat energy to make a small change in the temperature of water. Therefore water heaters are quite expensive to operate. It takes 1 BTU of heat energy to raise the temperature of 1 pound of water 1° Fahrenheit.

1.   Turn on a water faucet so that it is leaking at a rate of about one drip per second.
2.   Capture this water for a period of fifteen minutes.
3.   Weigh the amount of water you have collected (in pounds).
4.   Multiply this number by 35,040 to find the amount of water that would be lost in one year.
5.   Assume that water entering the water heater enters at 40° F and leaves the heater at 120° F.
6.   How many BTUs of heat energy would be lost in one year if the leak were not fixed?
      Number of BTUs/year = pounds of water in 15 min × 35,040 × 80 =
7.   How many gallons of fuel oil would it take to produce this much heat?
8.   Record results on data sheet.

Lighting

Lighting is something that we take for granted. We usually simply flip a switch and it is instantly there. However, what does it cost in energy to provide the light we use and does it make a difference what kind of light we use?

1.   In a dark room, hold a light meter exactly 10 feet from a 40-watt incandescent light bulb.
2.   Record the reading on the light meter.
3.   Similarly measure the amount of light coming from a 40-watt fluorescent light bulb.
4.   Since both bulbs have the same wattage, they use the same amount of electrical energy in the production of light. Which of the bulbs is more efficient in providing light?
5.   Approximately how much more efficient is this bulb?
6.   Record results on data sheet.

Transportation

North Americans look at freedom of movement as a right. We drive and fly more than any other people in the world. In some urban areas trains provide efficient ways to travel about the city. The millions of miles of highways make it easy for us to travel hundreds of miles. Airports are crowded to the point of being dangerous.

1.   For one week, keep a log of all the miles you travel by the following methods:
      a.   Foot
      b.   Bicycle
      c.   Automobile
      d.   Train or other rapid transit
      e.   Plane
      f.   Other
2.   Approximately what percentage of each of these was done just for fun?
      a.   Foot
      b.   Bicycle
      c.   Automobile
      d.   Train or other rapid transit
      e.   Plane
      f.   Other
3.   Record on data sheet.

Electrical Appliances

Electrical appliances are very convenient. They allow us to do things quickly and relieve us of distasteful or tedious tasks. How much energy do you use as a result of such devices? It is important to recognize that the total energy cost of an appliance also includes the energy necessary to manufacture, distribute, and merchandise the item. However, it might be instructive to determine how much energy is consumed by the use of the various electrical appliances we use. You can find the wattage of an electrical appliance on a label on the appliance.

1.   Keep a log of all the electrical appliances you use in a one-week period. List the appliance and the number of minutes it was used per week.
2.   Record the wattage of the appliance from the label on it.
3.   If you know the wattage and the number of minutes it was used you can calculate the number of kilowatt-hours of energy used (see the equation below).

$$\text{kilowatt-hours used} = \frac{(\text{watt rating})(\text{total minutes used}/60)}{1,000}$$

4.   If one kilowatt-hour is equal to 3,413 BTUs, how does energy consumption by using electrical appliances compare to energy consumption by automobiles or for home heating and cooling?
5.   Record results on data sheet.

Personal Energy Consumption
Data Sheet

Name _____

Section _____

Heating and Air Conditioning

Window	Surface area ft²	Inside temperature ° F	Outside temperature ° F	Temperature difference	R-value	Heat loss/year (BTU)
Single-pane					0.9	
Single-pane (inside temperature 5° F cooler)					0.9	
Single-pane (surface area ÷ 2)					0.9	
Double-pane					1.85	
Triple-pane					2.8	
Low-emission glass						

Heating Water

Number of pounds of water collected in 15 minutes	Amount of water lost per year	BTUs needed to heat water lost
	× 35,040 =	× 80 =

Gallons of fuel oil needed to heat the water lost: _____

Lighting

	40-watt incandescent	40-watt fluorescent
Light meter reading		

How much more efficient is the fluorescent bulb? _____

Transportation

1. For one week, keep a log of all the miles you travel by the following methods:
   a. Foot
   b. Bicycle
   c. Automobile
   d. Train or other rapid transit
   e. Plane
   f. Other
2. Approximately what percent of each of these was done just for fun?
   a. Foot
   b. Bicycle
   c. Automobile
   d. Train or other rapid transit
   e. Plane
   f. Other

Electric Appliances

Appliance	Wattage	Minutes used	Kilowatt hours used
Microwave oven			
Electric stove			
Hair dryer			
Stereo system			
Home computer			
Television			
Dishwasher			
Garbage disposal			
Electric shaver			
Space heater			
Electric fan			
Washer			
Electric blanket			
Vacuum cleaner			
Electric clothes dryer			

Energy Sources and Uses—A Comparison

Objectives

1. Consider alternative energy sources and efficient uses of energy.
2. Compare the energy efficiency of bicycles and cars.

Introduction

Because of the nonrenewable nature of oil, coal, natural gas, and nuclear energy, we will eventually need to develop other sources of energy. Energy conservation, using less energy, can be thought of as a source of energy just as a power plant would also be considered a source of energy.

Many studies have shown that energy conservation provides the greatest return for the dollar spent. It costs power companies less to promote energy conservation than to build new power plants. Generally, energy conservation programs also create more jobs.

Every time you rely on your body's energy to get you to work, school or the market, you and the environment get healthier. Riding a bicycle is more energy efficient than walking or riding in a car.

Procedure

In this exercise we will calculate the energy used to travel a given distance by bicycle and by car. We will use the resulting data to draw conclusions of energy efficiency and energy conservation.

Human-Powered Transportation versus Gas-Engine-Powered Transportation

As mentioned in the previous exercise, North Americans commute, travel, and generally move about the planet more than any other people. Note the number of miles you entered in your log in Exercise 22 for bicycle and automobile travel. We will calculate the energy use associated with traveling that total distance by bicycle and by automobile.

To calculate the energy required for your car to cover the distance recorded in your log, use the following two equations:

$$\text{gallons of gas used} = \left(\frac{\text{total miles traveled}}{\text{fuel efficiency (mpg)}} \right)$$

$$\text{total energy consumption} = \text{gallons of gas used} \times \left(\frac{125,000 \text{ BTUs}}{\text{gallon of gas}} \right)$$

Enter your hypothetical car-related energy consumption here: _____ BTUs

To calculate the energy use of using your bicycle to cover the same distance, first decide at what speed you usually ride your bicycle (15–20 m.p.h. is average). Use "Cal/mi" column of the table below to find your consumption of calories per mile of bicycle travel.

Human Energy Consumption While Bicycle Riding*

mph	Cal/mi	Cal/hr
10	26	260
15	31	465
20	38	760
25	47	1,175
30	59	1,770

*These values of efficiency, which were calculated from a bicycle efficiency modeling program, also yield rough equivalence to an informally given table of calorie consumption vs. velocity, appearing in Bicycling magazine around January 1990.

$$\text{total calories consumed} = \text{miles traveled} \times \left(\frac{\text{calories}}{\text{mile}}\right)$$

$$\text{total energy consumed (BTUs)} = \text{total calories consumed} \times \left(\frac{3.97 \text{ BTUs}}{\text{calorie}}\right)$$

Enter your hypothetical bicycle-related energy consumption here: _____ BTUs

Energy Sources and Uses—
A Comparison Data Sheet

Name _____

Section _____

1. What conclusions can be drawn about the energy consumed while riding a bicycle versus driving an automobile? In comparison, which form of transportation uses less energy?

2. Because automobile fuel is a nonrenewable form of energy (except in the case of gasohol derived from corn or sugarcane), alternative sources of automotive fuel will need to be developed. Is it possible that energy conserved through bicycle riding could eliminate the need for the development of such new fuels?

3. The social costs of driving an automobile are actually much higher than the out-of-pocket costs borne by each individual driver. This is because society must pay for roads, the ill effects of automotive pollution, the costs of pollution produced during production of new cars and disposal of old cars, the costs and damages associated with salting roads in the winter—the list goes on. These social costs are estimated to be in the range of $300 billion per year. Do you think these costs could be reduced by increasing the use of technologies such as bicycles?

4. Are bicycles and cars comparable goods—that is, can bicycling be considered a replacement or an equivalent good when compared to automobile travel? In what situation are the two forms of travel comparable (i.e., urban travel, rural travel, short distance, long distance, light loads, heavy loads, etc.)? In what situations would bicycling be considered substantially *not* a replacement for automobile travel?

The Effects of Radiation on the Germination and Growth of Radish Seeds

Objectives

1. Calculate the effect of different amounts of radiation on germination percentages.
2. Quantify the effect of different amounts of radiation on the growth of the root and shoot of irradiated radish seeds.
3. Observe the effect of radiation on the time of germination.

> **Safety Considerations:** There are no safety hazards in this exercise. Although the seeds have been irradiated, *they are not radioactive.*

Introduction

Gamma radiation is a form of energy similar to X rays, which, along with other forms of radiation, is emitted from a variety of environmental sources. Two other kinds of radiation are beta radiation and alpha radiation. Beta radiation consists of rapidly moving electrons, and alpha radiation consists of rapidly moving particles that are composed two protons and two neutrons. Atoms of the same element that differ from one another in the number of neutrons present are called *isotopes*. Some isotopes of atoms, such as cobalt 60, are natural sources of gamma radiation. X-ray machines and nuclear power reactors are also sources of gamma radiation. Radiation is of concern because, depending on its type, total amount, or rate of delivery, radiation can cause changes in the genetic material (DNA) within cells, or change the activities of cells. In very high doses, it kills cells directly.

Radiation dosage is measured in *rads*. A rad (*r*adiation *a*bsorbed *d*ose) is equal to 100 ergs of energy absorbed by 1 gram of material. Most human radiation doses are measured in millirads (mrad = 1/1,000 of a rad). Another unit that is often used to quantify radiation dosages is the *rem* (*r*oentgen *e*quivalent *m*an). It is a slightly different way of measuring radiation, but for gamma radiation the rad and rem are equivalent.

The radish seeds that you will examine in lab have received the following amounts of radiation from a cobalt 60 source:

Unirradiated control
Irradiated 50,000 rads
Irradiated 150,000 rads
Irradiated 500,000 rads
Irradiated 4,000,000 rads

The following chart gives some idea of the amounts of radiation organisms are likely to encounter.

The Effects of Some Common Radiation Exposures

Kind of exposure	Dose (amount of exposure)	Effects of exposure (at dose shown)*
Color television set	0.5 mrad/yr	None known
Upper limit allowable from a nuclear power facility	5 mrad/yr	None known
Chest X ray	5.8 mrad/X ray	None known
Brain scan X ray	100 mrad/X ray	None known
Natural background	100–200 mrad/yr	None known
Allowable maximum radiation exposure for general public	500 mrad/yr	None known
Barium enema X ray	800–1,500 mrad/X ray	None known
Allowable exposure for a worker in a nuclear power plant	5,000 mrad/yr	None known
Radiation treatment	10,000 mrad/yr	Early embryo abnormal
Lowest exposure known to cause leukemia in survivors of Hiroshima	20,000 mrad/yr exposure	Leukemia

*There is strong evidence that the effects of radiation can be cumulative. Therefore a single small dose may not in itself be harmful, but if it is added to lots of other small doses over a lifetime there may be effects that show up late in life.

Procedures

1. Place ten of each kind of radish seed on wet paper towels in a petri dish.
2. Record the number germinated in each dish every eight hours.
3. Record the total number germinated after two days.
4. Determine the average length of the roots and shoots at each radiation dose after two days.
5. Answer questions on the data sheet.

Directions

1. Obtain five petri dishes with lids, which contain wet paper towels cut to fit the petri dish.
2. Use a wax pencil or piece of tape to label the petri dishes as follows (label the bottom with the seeds in it, not the cover, to help prevent misidentification when the seeds are examined):

    control
    50,000 rads
    150,000 rads
    500,000 rads
    4,000,000 rads
3. Place ten seeds of the appropriate kind in each of the petri dishes. They should be spread out on the paper towel. They should begin to germinate in about eight hours. Germinated seeds will have split the brownish seed coat, and you should be able to see a whitish or yellowish part of the young plant protruding from the seed coat.
4. Examine the seeds at regular intervals (every eight hours or at times specified by your instructor) and record the number of each kind of seed that has germinated on table 1.

Table 1   Number Germinated

Time	8 Hours	16 Hours	24 Hours	32 Hours	40 Hours
Control					
50,000 RADS					
150,000 RADS					
500,000 RADS					
4,000,000 RADS					

5. After two days or a period of time specified by your instructor, record the total number germinated and the average length of the roots and shoots on table 2. You will need to remove the plants from the petri dishes and use a millimeter ruler to do this. Since some roots may be curved, you will need to take this into account when measuring them.

Table 2   Data on Growth

	Percentage germinated	Average length of root (mm)	Average length of shoot (mm)
Control			
50,000 RADS			
150,000 RADS			
500,000 RADS			
4,000,000 RADS			

6. On the data sheet, construct graph 1, showing the percentage of the seeds that germinated versus time. You will have five different lines on this graph, so make sure that you label each.
7. Combine all of the table 1 and 2 data from all students in class and construct graphs of germination percentages versus radiation dose (graph 2), average root lengths versus radiation dose (graph 3), and average shoot lengths versus radiation dose (graph 4). Because there is a wide range of dosages, it is convenient to plot the log of the dosage.

The Effects of Radiation on the Germination and Growth of Radish Seeds Data Sheet

Name _____

Section _____

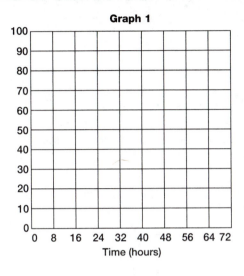

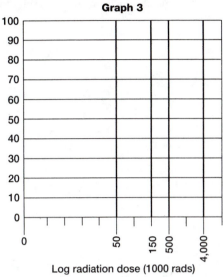

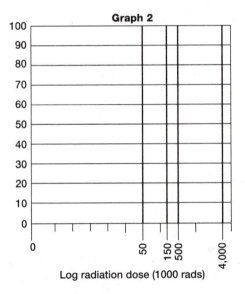

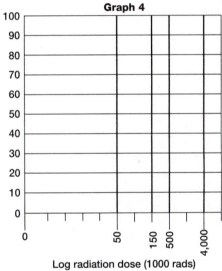

1. Is there a dose at which no effect on time of germination is noted?

2. What dose caused the greatest reduction in germination percentage?

3. What dose caused the greatest change in root length?

4.   What dose caused the greatest change in shoot length?

5.   Are there any dosages at which the times of germination, total percentage germinated, average root lengths, or average shoot lengths do not differ from the control? Explain why this might be significant.

6.   As a result of this exercise, what do you think would happen to plants in the vicinity of a nuclear accident such as happened in Chernobyl in the former Soviet Union?

Objectives

1.   Become familiar with renewable and nonrenewable energy options by participating in a classroom debate.
2.   Understand the process involved in coming to a group consensus on an issue.
3.   Take steps to formally voice your opinion in the form of a letter.

Introduction

Worldwide consumption of nonrenewable resources is proceeding at such a rate that consumption patterns will be forced to change as resources are depleted unless steps are taken to reduce resource use by initiating proactive resource management and conservation strategies. Energy resources fit within this category. In one year, worldwide consumption of fossil fuels (coal, oil, gas, natural gas) is equal to the amount of fossil fuels created over one million years through natural processes.

In addition, energy use is associated with air, water, and land pollution. In the case of air pollution, compounds released into the air during fossil fuel use have been linked to global warming trends, urban photochemical smog, and ozone depletion. In the case of nuclear energy, extremely long-term waste management strategies must be developed and implemented.

Renewable energy sources such as solar energy, wind, and hydroelectric power avoid some of the pollution and waste management problems posed by nuclear and fossil fuel consumption. Renewable resources such as firewood have effects similar to fossil fuels, and their use can result in the destruction of forest ecosystems.

Procedures

With the knowledge you have gained in this class and through everyday experiences, you will be asked to participate in a debate. Your instructor will divide the class into two groups. One group will argue for renewable energy resources, the other will advocate the use of nonrenewable resources. You will be placed in a group regardless of your beliefs for or against a particular energy source. The intent of this exercise is to bring to your attention new ideas and stimulate exploration of different approaches to energy consumption and energy source issues. Your instructor will act as mediator for this exercise, correcting any nonfactual information that may arise during discussion.

During the debate, you may wish to discuss issues such as the relative cost of different energy technologies, the differences of short-and long-term perspectives, the effects of energy-use taxes on energy consumption patterns, and so on.

Upon completion of the debate, you will be responsible for compiling a list of the pros and cons of each energy source. When this list is complete, return to the debate. The objective now is to come to a group consensus on the best ways to manage energy use. What sources of energy will be emphasized? What, if any taxes should be introduced? What incentives should be introduced to alter consumption patterns? What steps would need to be taken to implement this plan?

What effect might these policies have on the national economy? Will certain industries be hurt? Will other industries be helped? What effects on the quality of life might result?

Finally, each member of the class is required to compose two letters. One letter will be to a local government official or group of officials. Another letter will be to a federal government official. The letters will state your concerns about energy sources and uses and suggest steps you think government needs to take to help resolve the issues.

Do not hesitate to write additional letters. The more letters you write, the better the chance of your point of view being accepted. Invite friends and family to participate as well.

Energy Debate
Data Sheet

Name _____

Section _____

1. During the debate, was it difficult to accommodate everyone's belief into **one** solution? Explain. Does this help to explain the position of government? If you had difficulty getting your class to reach an acceptable consensus, imagine trying to consider the opinions of millions or billions of individuals!

2. Which energy source do you feel has the most detrimental affect if used heavily? Which energy source is the most dangerous if used heavily? Which energy source is the most beneficial? Which is the safest? Explain.

3. What affect do you expect your letters to have? How many letters do you think a U.S. Congressperson receives each day? How are these letters used in decision making?

4. What types of obstacles must be overcome if the changes you recommend in your letter are to be implemented? What affects will the changes you recommend have? Generally, what classes of political action special interest groups will likely be in favor of your plan? Which will oppose your plan?

Field Trip Suggestions

1. Visit a local industry's water pollution-control facilities. Determine the cost of operating the facility. What are the pollution problems the industry is trying to control? What techniques are used?
2. Visit an orchard, cotton field, or vegetable farm. Determine which pesticides are used and what the purpose of each pesticide is. Determine the cost of the pesticides used each year. Determine what special training is required for pesticide applicators. Determine the conditions that determine when pesticides are applied.
3. Canoe down a local stream or river through an industrial area and record examples of pollution. Consult with your local pollution-control agency about your impressions. Do they agree or disagree that the examples you cite are important?
4. Fly over or drive through an industrial area and record examples of pollution. Consult with your local pollution-control agency about your impressions. Do they agree or disagree that the examples you cite are important?
5. Visit your local solid waste disposal site. How many tons of material are buried per day?

Alternative Learning Activities

1. Invite a spokesperson from an environmental group to talk to the class about the major pollution problems in your area.
2. Collect newspaper articles that relate to pollution. Note the sources used in writing the article. If the article is by a local writer, phone him or her and try to determine what he or she knows about the subject, or ask the writer to come to class to discuss how information was gathered for the article. Choose five articles and comment on the validity of the impressions they give.
3. Have a person from your local extension service office come to the class to discuss the pros and cons of pesticide use and the licensing requirements for pesticide applicators in your state.
4. Have a spokesperson from a local pollution-control agency visit the class and discuss its function.
5. Do an inventory of hazardous chemicals in your home. Determine the proper disposal method for each item found.
6. Draw a diagram tracing the flow of wastewater from your toilet to its discharge into a local body of water.
7. Draw air through a filter. Examine the filter under a microscope. See if you can identify the sources of any of the particles you find.
8. Sit quietly in a room, park, or natural area. Identify the sounds you hear. Rank them as to which are the loudest and which are the most annoying.
9. Keep a detailed diary of the amount of solid waste you produce in one day. List ten ways you could reduce the amount of waste you produce.
10. Read *The Tragedy of the Commons* by Garrett Hardin and apply his concept to a local situation. Write a paragraph showing how the local situation is related to Hardin's essay.
11. Capture rain or snow and measure its pH.
12. Maintain several bottles of pond water in sunlight. Add different amounts of fertilizer to each bottle and compare how they differ in appearance over time.

Objectives

1. Understand the relationship between air pollution and the combustion of various materials.
2. Better understand how society and industry add potentially harmful pollutants to the air.
3. See the difference between the air pollutants given off from the combustion of natural and of synthetic materials.
4. More fully understand that the individual is a source of air pollution.
5. Better appreciate that, by altering consumer buying practices, air pollution can be reduced.

Safety Considerations: When burning various materials, make sure there is adequate ventilation and observe fire safety laws.

Introduction

Five major types of materials are released into the atmosphere in sufficient quantities that, in their unmodified form, they are considered to be primary pollutants. Carbon monoxide (CO), a colorless gas, is produced when organic material, such as gasoline, coal, wood, and trash is incompletely burned. Smoking tobacco also produces large amounts of CO. In addition to CO, automobiles emit a variety of hydrocarbons (HC). Hydrocarbons are either evaporated from fuel supplies or are remnants of the fuel that did not burn. Particulates constitute the third largest category of air pollutants. Particulates frequently receive a great deal of public attention because they are so readily visible. An example is the black smoke coming from a factory. Sulfur dioxide (SO_2) is a compound containing sulfur and oxygen that is produced when sulfur-containing fossil fuels are burned. SO_2 has a sharp odor and irritates respiratory tissue. Oxides of nitrogen (NO and NO_2) are the fifth type of primary air pollutants. Nitrogen compounds produce a reddish brown color in the atmosphere and react with other compounds to produce photochemical smog.

As you can see, air pollution varies considerably. Air pollution can be detected and in some cases diagnosed by using your senses of sight, smell, and touch. In this exercise you will produce various primary air pollutants and observe their by-products.

Procedures

1. Place various materials under a beaker and ignite them.
2. After each material has burned, observe the products of combustion.
3. Record observations.
4. Based on the observations, respond to the questions in the data sheet.

Laboratory Exercise

1. Place a small wad of cotton on a jar lid and place the lid on a clean half-sheet of paper. Label the paper *cotton* (see diagram).
2. Divide into groups. Using a match, one student should light the cotton and cover the lid with a 600 ml beaker. If the flame begins to go out before the cotton is fully burned, lift the edge of the beaker slightly to let in more air. Observe the products of combustion.
3. When the flame goes out, study the beaker and its contents for several minutes. Record observations on the air-pollution data sheet at the end of this exercise. Take the beaker to a window and allow fumes to escape.
4. Use a fresh wad of cotton, clean paper, and a clean beaker. With a medicine dropper, place two drops of turpentine on the cotton. Repeat the procedures for burning and observation discussed in steps 1–3. Enter your observations on the air-pollution data sheet.
5. Repeat the same procedures for the burning of wool and Styrofoam, and enter the data in the table.
6. Repeat the same procedures with other material suggested by your instructor.

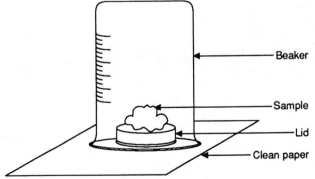

Beaker

Sample

Lid

Clean paper

Air Pollution
Data Sheet

Name _____

Section _____

Material	Color of smoke	Odor of smoke burning	Other observed products	Other observations
Cotton				
Turpentine and cotton				
Wool				
Styrofoam				

1. Which of the materials gave off the most air pollution? On what do you base your answer?

2. What substance when burned produced liquid inside the beaker? What color was the liquid?

3. Did you find any solid pollutants being given off by the combustion of these materials?

4. Which of the products—synthetic or natural—seemed to give off more pollutants in solid and gaseous form? What additional data or experiments would you need in order to answer this question completely?

5. What pollutants do you think you have added to the air so far today? Can you think of alternative ways of carrying out the day's activities so you pollute the air less?

6. What potentially harmful pollutants are added by autos, coal power plants, nuclear power plants? What are some alternatives to these activities that might produce fewer air pollutants?

7. What effect do you feel synergism has in air pollution?

Objectives

1. Increase awareness of the problems caused by acid deposition.
2. Determine the acidity of various samples of water in the community and draw comparisons between the acidity of samples and deposition.

Safety Considerations: When collecting water samples from area watercourses, observe basic safety considerations.

Introduction

The acidity of a liquid is determined by its concentration of hydrogen ions (H^+). This concentration is described using a pH scale (figure 27.1). The scale ranges from 0.0 (most acidic, highest H^+ concentration) to 14.0 (most basic, lowest H^+ concentration). A solution with a pH of 7.0 is neutral. The pH scale is a negative logarithmic scale in base 10. This means that a solution with a pH of 4.0 is ten times more acidic than a solution with a pH of 5.0 and 100 times more acidic than a solution with a pH of 6.0. Normal precipitation has a pH of about 5.6. Acid rain is defined as having a pH of 5.0 or less.

Acid rain is becoming a worldwide problem. During the past decade, increasing attention has been paid to the human activities that produce sulfur dioxide (SO_2) and oxides of nitrogen (NO_X). It is estimated that 99% of all SO_2 comes from the internal combustion engine and the use of coal for industrial operation and power generation. Oxidizing agents, such as ozone, hydroxyl ions, or hydrogen peroxide, along with water, are necessary to convert the sulfur dioxide or nitrogen oxides to sulfuric or nitric acid. This acid is washed from the air when it rains or snows. The acid damages plants and animals and increases corrosion of building materials and metal surfaces.

Procedures

1. Collect water samples from various sources, including the following: rain, snow (if possible), nearby streams, lakes, bogs, or ponds over a period of time.
2. Measure the pH of the samples.
3. Develop graphs or charts of data obtained.

Collection and Analysis of Water Samples

1. Using clean glass, collect rain or snow samples from campus and/or home. Note where samples were collected. Runoff from roofs or trees may have altered pH. Using pH paper, measure the pH of each sample. Record the data on the data sheet. As a term project, you may wish to collect samples over a period of several weeks or months.
2. Take a field trip to collect water samples from nearby streams, lakes, bogs, or ponds. You may also collect water from alternative sources, such as water from the tap, drains, fountains. Use distilled water as a control. Be sure to label the samples. If possible, collect several samples from each location so that you can obtain an average pH reading. Compare the pH's of the samples and record them on the data sheet. What might the pH tell about what aquatic organisms can live in the water from the waters sampled?
3. Compare the pH of samples collected at different locations during the same storm, or of samples collected at different times at the same location, and answer the following questions:
   a. Do the pH's differ at all?
   b. What are the possible explanations for similarities or differences? For example, how might the path of a storm affect the pH of its precipitation?
   c. Check weather reports and maps to track the path of the storm. Remember, normal pH of rain is 5.6.
4. After analyzing the samples, develop graphs or charts of the data in class and answer the following questions:
   a. What effect does pH have on water and on living things?
   b. Identify local and distant sources of pollutants that affect pH.
   c. What do you think should be done to reduce the acid rain problem in North America?

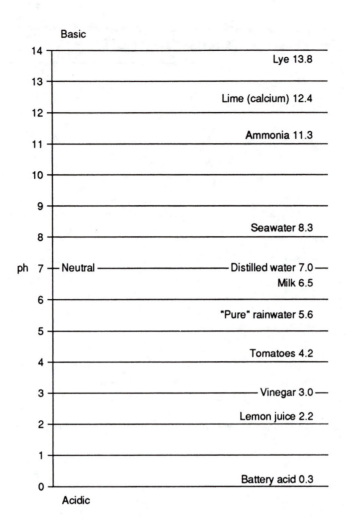

Figure 27.1   pH scale.

Acid Deposition Sampling Data Sheet

Name _____

Section _____

Sample	Date collected	Location	Type of precipitation	pH
1				
2				
3				
4				
5				
6				
7				
8				
9				
10				
11				

Objectives

1. Measure the effect of various toxic materials on brine shrimp (*Artemia salina*).
2. Determine the LD50 (Lethal Dose 50%) for a variety of toxic materials.

Safety Considerations: Some of the materials used are toxic to humans as well as to brine shrimp. Handle these materials so that they do not come in contact with your skin.

Introduction

We handle many materials daily that are toxic. We are often unaware of the degree to which they are toxic. In this laboratory we will look at water solutions of several household materials and determine their toxicity to brine shrimp, *Artemia salina.*

Measuring toxicity requires a standard method of comparison. A typical method is to determine the concentration of a toxic material that is able to cause 50% mortality in a population of test animals. This is called an LD50 (Lethal Dose 50%) test of toxicity. For a variety of reasons, different animals respond differently to the same toxin. Some animals may be very sensitive to a toxin, whereas others are relatively resistent to its effects. Because species of animals vary, it is important to understand that what is toxic to brine shrimp may not necessarily be toxic to other kinds of animals to the same extent.

Many household items that we deal with on a regular basis are toxic materials, but we don't usually think of them as being toxic. It should be instructive to examine several such materials to determine their toxicity.

Procedures

During this exercise you will determine the LD50 for $CuSO_4$ and several common household chemicals. Your instructor may want to divide the class into groups so that more kinds of materials can be tested.

Directions

1. Obtain five petri dishes.
2. Label these dishes as follows:
    10% $CuSO_4$
    1% $CuSO_4$
    0.1% $CuSO_4$
    0.01% $CuSO_4$
    0% $CuSO_4$ control
3. Place 10 ml of the solution indicated in the respective petri dishes.
4. Use an eyedropper to place ten brine shrimp in each of the dishes.
5. Record the date and the time.
6. Examine the petri dishes at twenty-four hours and forty-eight hours and record the number of brine shrimp that have died on the chart on the data sheet.
7.   Plot your data on the data sheet graph and determine the approximate concentration at which 50% of the brine shrimp died at the end of forty-eight hours. This is the LD50 for $CuSO_4$.
8. Determine the LD50 for two other common household chemicals. Choose from the following list, or select others with the help of your instructor. Possible choices are vinegar, aspirin, rubbing alcohol, cold coffee, nicotine, and hydrogen peroxide. Use a wide range of concentrations, from very dilute to very concentrated. Your instructor will provide you with assistance in determining concentrations. Record data on data sheet.

Toxicity Testing
Data Sheet

Name _____

Section _____

Data for CuSO₄

Substance concentration	Number of brine shrimp dead	
	24 hours	48 hours
10% CuSO₄		
1% CuSO₄		
0.1% CuSO₄		
0.01% CuSO₄		
Control 0% CuSO₄		

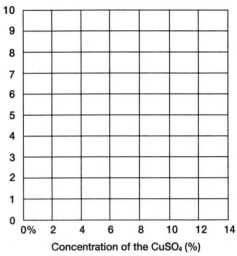

Number of dead organisms

Concentration of the CuSO₄ (%)

Data for Household Substance 1

Substance concentration	Number of brine shrimp dead	
	24 hours	48 hours

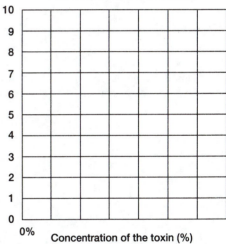

Number of dead organisms

Concentration of the toxin (%)
Name of toxin: _____

Data for Household Substance 2

Substance concentration	Number of brine shrimp dead	
	24 hours	48 hours

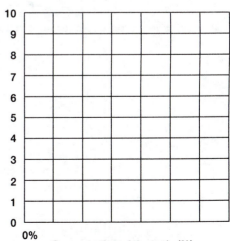

Number of dead organisms

Concentration of the toxin (%)
Name of toxin: _____

1. Which of the substances tested was the most toxic?

2. LD50 tests have been called inhumane. Should they be allowed? Why or why not?

3. Would you feel differently if the animals being tested were mice? Why?

4. How do the results you obtain relate to humans? Do you think they would have the same level of toxicity to humans? Could humans be poisoned by aspirin, rubbing alcohol, or vinegar?

5. What is the purpose of the control in each experiment?

Dissolved Oxygen

Objectives

1. Understand the relationship between the presence of oxygen in water and the overall quality of the water.
2. Observe the effects of nutrients and temperature on levels of dissolved oxygen.

Safety Considerations: Wash your hands after contact with the water after sampling. Use protective eyewear when working with the chemicals in the dissolved oxygen test kit. Some of the chemicals used may stain clothing.

Introduction

Dissolved oxygen is critical for the maintenance of environmentally healthy rivers and lakes. Oxygen is important in water because it is used by organisms for respiration and to break down organic compounds. The presence of oxygen in water is a positive signal, whereas the absence of oxygen is a signal of environmentally degraded water. The amount of oxygen in the water is influenced by several factors such as temperature, turbulence, and photosynthesis. Human impacts, however, have become a major factor contributing to changes in dissolved oxygen levels in water. Organic wastes from human sewage and industries are a primary source of the problem. Agricultural runoff of fertilizers is also a major problem. As the runoff enters the water it stimulates the growth of algae and other aquatic plants. As the plants die, aerobic bacteria consume oxygen in the process of decomposition. Many kinds of bacteria also consume oxygen while decomposing sewage and other organic matter in the water.

Dissolved oxygen in water is measured in parts of oxygen per million parts of water (ppm). Different organisms require different amounts of dissolved oxygen in order to survive. Trout, for example, need roughly 6.5 ppm, carp need 2.5 ppm, and sludge worms can live in 0 ppm of dissolved oxygen.

The effects of oxygen-demanding wastes on rivers depend to a great extent on the volume, flow, and temperature of the river water. Aeration occurs readily in a turbulent, rapidly flowing river, which is therefore often able to recover quickly from oxygen-depleting processes. Downstream from a point (such as a municipal sewage plant discharge), a characteristic decline and restoration of water quality can be detected either by measuring dissolved oxygen content or by observing the flora and fauna that live in successive sections of the river.

The oxygen decline and rise downstream are called the **oxygen sag** (see figure 29.1). Above the pollution source, oxygen levels support normal populations of clean-water organisms. Immediately below the source of pollution, oxygen levels begin to fall as decomposers metabolize waste materials. Rough fish, such as carp, bullheads, and gar, are able to survive in this oxygen-poor environment, where they eat both decomposer organisms and the waste itself. Farther downstream, the water may become anaerobic (without oxygen), so that only the most resistant microorganisms and invertebrates can survive. Eventually most of the nutrients are used up, decomposer populations become smaller, and the water becomes oxygenated once again. Depending on the volumes and flow rates of the effluent plume and the river receiving it, normal communities may not appear for several miles downstream.

Procedures

1. Obtain a dissolved oxygen test kit.
2. Take water samples from several sites. The sites should be areas of differing water quality (i.e., a clean river to a more visibly polluted pond or industrial area).
3. Using your dissolved oxygen test kit, measure the dissolved oxygen level of your samples. Take the water temperature at each site.
4. Calculate the dissolved oxygen saturation percentage for each sample using the chart provided in this exercise.
5. Answer the questions in the data sheet.

Experiment: Measuring the Dissolved Oxygen in Water Samples

Dissolved oxygen levels in water vary according to weather, time of day, and temperature. In addition it is best to sample water away from shore and below the surface. All tests for dissolved oxygen should be run immediately after sampling. Choose several local sites to investigate for dissolved oxygen and record percent saturation on the data sheet.

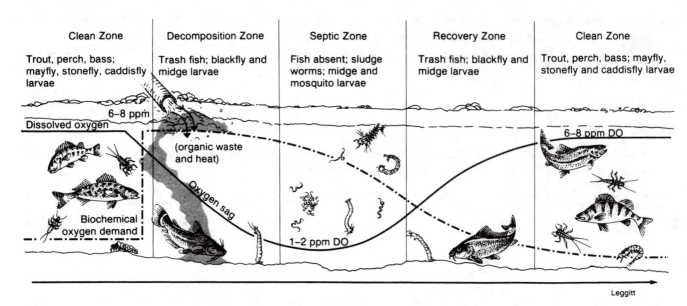

Figure 29.1   Oxygen sag downstream of an organic source. A great deal of time and distance may be required for the stream and its inhabitants to recover. (From William Cunningham and Barbara Saigo, *Environmental Science: A Global Concern,* 2d ed. Copyright © 1992 Wm. C. Brown Communications, Inc., Dubuque, Iowa. All Rights Reserved. Reprinted by permission.)

Procedures

Your dissolved oxygen test kit should contain the following:

1. Bottle with stopper
2. Dissolved oxygen reagent pillow #1
3. Dissolved oxygen reagent pillow #2
4. Clippers
5. Phenylarsine oxide solution (PAO)
6. Measuring tube
7. Square mixing bottle
8. Eyedropper
9. Dissolved oxygen pillow #3

Determining Dissolved Oxygen

1. Remove the stopper and immerse the DO bottle beneath the water surface.
2. Allow the water to overflow for two to three minutes (this will ensure the withdrawal of air bubbles).
3. Add the contents of pillow #1 (manganous sulfate powder) and pillow #2 (alkaline iodide azide powder) to the DO bottle.
4. Insert the stopper, making sure no air is trapped inside, and shake.
5. Allow the sample to stand until the precipitate settles halfway. When the top half of the sample turns clear, shake again, and wait for the same changes.
6. Add pillow #3 (sulfamic acid powder) to sample and shake. The precipitate will dissolve and the water will turn yellow.
7. Pour sample to the top of the measuring tube; pour the contents of the measuring tube into the square mixing bottle.
8. While swirling the sample to mix, add PAO Titrant to the prepared sample one drop at a time. Count the number of drops needed to change the sample yellow color to a clear solution. Hold the bottle against white paper to see the color change accurately. Each drop equals 1.0 mg/liter of dissolved oxygen.

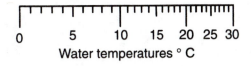

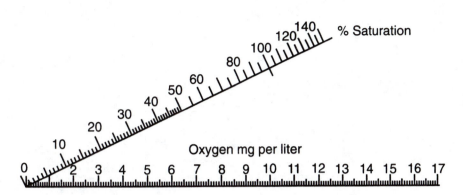

Figure 29.2   Level of oxygen saturation chart.

Calculating Percent Saturation

The percent saturation of dissolved oxygen present in water at a given temperature is determined by pairing temperature of the water with the dissolved oxygen value. It is determined through the use of the percent saturation chart.

To use the chart (figure 29.2), draw a straight line between the water temperature at the test site and the dissolved oxygen measurement (oxygen mg per liter) and read the saturation percentage at the intercept on the sloping scale.

Example: Let's say that your dissolved oxygen value was 8 mg/l and the measured water temperature was 15° C. Drawing a straight line between these two values gives a percent saturation of about 80%. That figure means that at 15° C you would expect a higher dissolved oxygen value—something is using up oxygen in the water.

Dissolved Oxygen
Data Sheet

Name _____

Section _____

Site	Dissolved oxygen (ppm)	Percent saturation

1.   Did you identify areas that had abnormally low dissolved oxygen?

2.   What effect does the temperature of the water have on dissolved oxygen?

3.   List several factors that could lower oxygen concentration.

Objectives

1. Understand the relationship between water quality and total solids.
2. Better understand the human-caused sources of total solids in a watercourse.
3. Better understand how levels of human-caused suspended solids can be reduced.

Safety Considerations: Use gloves when obtaining water from contaminated sites.

Introduction

Dissolved inorganic materials occur naturally in water. Calcium, bicarbonate, phosphorus, nitrogen, sulfur, iron, and other ions are found in water bodies. These materials are necessary for aquatic life because the overall concentration of total solids determines the diffusion of water in and out of the cells of organisms.

Total solids include suspended solids as well as dissolved solids. Suspended solids include clay and silt particles from soil erosion, particulate matter from sewage, and industrial wastes. Suspended solids can come from natural occurrences such as floods; however, there is usually a human factor involved.

High concentrations of total solids may cause lower water quality and threaten individual organisms. A high concentration of suspended solids reduces water clarity and decreases photosynthesis. In addition suspended solids may bind with toxic compounds and heavy metals, allowing them to become available to be ingested by animals. Suspended solids may also increase water temperature through greater absorption of sunlight. Farming practices, urban land use, road salt, waste from sewage treatment facilities, and residential use of fertilizers all contribute to the problem of total solids.

Procedures

1. Use a glass-stoppered bottle to obtain a water sample of at least 100 ml. It is best to obtain the sample away from shore and below the surface of the water.
2. In the laboratory clean a 300 ml beaker and dry it for one hour in an oven at 103° C. You may use a burner on low heat if you do not have an oven.
3. Remove the beaker from the heat and allow it to cool. Using a sensitive balance (to the nearest .0001 gram), weigh the beaker. Do not touch the beaker with your hands.
4. Transfer a 100 ml sample of the water into the 300 ml beaker.
5. Evaporate the 100 ml sample by putting the beaker in an oven overnight at 103° C. Once the beaker cools reweigh it.
6. Subtract the weight of the beaker (before the water sample was taken) from the weight of the beaker and residue to obtain the increase in weight, or the weight of the residue in grams. The formula for determining total solids is

$$\frac{\text{increase in weight in gm}}{\text{volume in milliliters (ml)}} \times \frac{1{,}000 \text{ mg}}{1 \text{ gram}} \times \frac{1{,}000 \text{ ml}}{1 \text{ liter}} = \text{mg/l}$$

Example: Weight of beaker and residue = 48.2982 grams

Weight of empty beaker	48.2540 grams
Weight of residue	.0442 grams

$$\frac{.0442}{100 \text{ ml}} \times \frac{1{,}000 \text{ mg}}{1 \text{ g}} \times \frac{1{,}000 \text{ ml}}{1 \text{ l}} = 442 \text{ mg/l}$$

7. Repeat the procedure using water from different sites: a rural stream, a water sample taken from a river or stream after a heavy rainfall, a water sample from an area under development. Record your findings on the data sheet.

Total Solids
Data Sheet

Name _____

Section _____

	Water sample	Total solids
Site 1		
Site 2		
Site 3		

1. What accounts for the different amount of total solids in the samples?

2. What are the major sources of suspended solids in your area? Is it possible to control or reduce the sources of total solids in your area?

Field Trip Suggestions

1. Visit your state, county, or city officials and discuss environmental legislation with them. Ask them to state their position on a locally important environmental issue. Ask them to state what environmentally significant legislation or ordinances they have supported in the past year.
2. Visit a local recycling center and list the kinds of materials recycled. Describe how the materials are processed within the center. Determine the market for the materials. Who uses the recycled materials, and how much are they willing to pay?
3. Visit your local water treatment plant. Make a list of the treatment methods employed and what purpose each method accomplishes. Ask what the major water quality problems are in your locality.
4. Visit a junkyard. Make a list of the kinds of materials accepted. Determine the price paid for each kind of material.
5. Attend a public hearing on an environmental issue.

Alternative Learning Activities

1. Invite your state or federal representatives or senators to address the class or school on environmental issues.
2. Write a letter outlining your position on a piece of environmental legislation to an appropriate government official.
3. Volunteer your time to participate in an environmentally significant activity.
4. Select a piece of land and do an environmental history of it. You may need to interview the current owners and search the records in the county register of deeds office.
5. Select a large piece of land and develop a map that shows suitable uses for various portions of it.
6. Determine what current environmentally significant legislation is before Congress (select one bill). How would passage of the bill affect your area? Keep a log that lists the progress of the bill.
7. Ask a spokesperson from your local water planning board to talk to the class about where water is used and how water use will affect the economy of the area.
8. Use a map of your local community. Locate all the open space that is available to the public (playgrounds, parks, golf courses, natural areas, and so forth). Do all regions of your community have equal access to open space? How does this relate to economic conditions, politics, and community planning?
9. Identify three places in your community that illustrate good planning and three that illustrate poor planning. Prepare visual aids that illustrate your points.
10. Determine how much water you use per day.
11. Pick up litter along a road or section of beach. Inventory the waste found. Identify causes of the litter and discuss solutions to the problem.

Objectives

1. Understand that land-use decisions may not always be carefully thought through.
2. Recognize that most types of land use may have unwanted or negative environmental results.
3. Understand the relationship between the development of shopping centers and the decline of central business districts.

Safety Considerations: There are no specific safety considerations in this exercise.

Introduction

Shopping centers are frequently constructed with inadequate total planning. Communities are quick to see the potential positive economic spinoff of a shopping center being developed but tend to downplay potential negative impacts of such development. Environmental considerations are not always properly addressed and thus surface later as problems. To understand what environmental problems may arise from the construction of a shopping center, it is necessary to analyze an existing or proposed facility.

Procedures

1. Design a questionnaire to be administered to neighboring residents of a local shopping center.
2. Administer the questionnaire.
3. Undertake an on-site evaluation of the shopping center.
4. Discuss and evaluate the data compiled from the questionnaire and the on-site evaluation.

Questionnaire Design

In developing the questionnaire, the class should address such questions as the following:
1. Did you live in this area before the shopping center was constructed? If so, what was the area like before the shopping center was built?
2. What problems do you notice associated with the shopping center (noise, traffic, etc.)?
3. What do you like about having a shopping center nearby?
4. Is the shopping center aesthetically pleasing (interior, exterior)?
5. Do you feel crime in your neighborhood has changed with the construction of the shopping center?
6. Were you consulted in any way before the shopping center was built? If so, in what way and by whom?

Shopping Center Evaluation

The evaluation conducted by the students should include such information as the following:
1. An estimate of the total area of impervious surfaces (any surface where water cannot seep through, such as buildings and parking lots)
2. The number and average sizes of trees in the shopping center
3. How the trees or shrubs are planted: in boxes, in areas filled over with wood chips, or in beds
4. An estimate of the parking area used at one time
5. A survey of the drainage, sewers, ditches, holding ponds, or nearby stream. Where does all the water that falls on the impervious surfaces drain?
6. Is the area aesthetically pleasing or not, and why?
7. Were the roads leading to the shopping center there before construction, or were they widened or constructed because of the shopping center?
8. What kinds of stores are at the shopping center? Are there any small shops or unique stores, or just large chain stores?
9. Is there evidence that the shopping center was constructed where a farm, residential area, or open fields used to be?
10. Is there evidence that large trees were removed for the shopping center? If so, how could they have been incorporated into the shopping center complex rather than having been removed?

Land-Use Planning: A Shopping Center
Data Sheet

Based on the data collected and the evaluation, answer the following questions. You may use additional information in your analysis.

1.   What functions do shopping centers play in your community?

2.   Are several small shopping centers better than one large one? Why or why not?

3.   Do zoning practices change when a shopping center is built, such as zoning from residential only to commercial or industrial?

4.   How does the presence of a shopping center affect the neighboring area? Do more businesses, apartments, or gas stations come into the area?

5.   What role does the construction of shopping centers play in the decay of the central business district of the city, the downtown?

6.   Are the environmental problems associated with large shopping centers inevitable? How could the problems be minimized?

Objectives

1. Help create an understanding that both the problems and the solutions associated with solid waste ultimately rest with the individual.
2. Increase an awareness of how life-style has a direct impact on the volume of solid waste generated.
3. Bring the problem of an ever-growing volume of solid waste down to the individual student life-style.
4. View alternatives at the individual level to help reduce the volume of solid waste generated.

Safety Considerations: Sanitary precautions should be observed when handling waste products.

Introduction

The ever-growing volume of solid waste and the corresponding need to dispose of it are problems that will reach crisis proportions in the near future. In many communities the problem has already become acute, forcing the enactment of mandatory recycling programs and expensive disposal costs. For most of us in North America, however, the problems associated with solid wastes appear, at least on the surface, rather removed. Solid wastes (or if you prefer, garbage) is not, after all, a subject most people like to think about. If they do, then the primary problem with it is hauling it to the curbside once a week, where it disappears. "Out of sight, out of mind" is a cliche that seems highly appropriate for solid waste. In fact, each of us contributes roughly 2 kilograms daily to the problem, or over 650 kilograms per year. In the United States alone, this amounts to 250 million metric tons of solid waste that must be buried, burned, or in some other way disposed of. For many areas, the bottom line is that there simply is no place to dispose of the growing volume of waste.

Where does all this waste come from? Must we wait until a major crisis develops before we attempt to address the problem? To a large extent the solution lies with you and your willingness to reduce the volume of waste *you* generate. One thing, however, is certain; if we do not voluntarily begin to reduce the amount of waste generated, we will all pay the price of dealing with the problem. This will be in increased product costs, increased taxes, and many new governmentally mandated rules and regulations. The problem must be and will be dealt with. It's really just a matter of how.

Procedures

1. Keep a record of all the solid waste you dispose of over a three-day period.
2. Record the type of waste on the data sheet at the end of this exercise. You will need to use additional data sheets.
3. Answer the questions on the data sheet.

Directions

1. Over the course of at least three days, keep an inventory of all items you dispose of. Record the items on the data sheet by material and weigh your accumulation of each kind of material.
2. Once you have completed your inventory, describe where your waste goes after it leaves your home or dorm. Describe each step involved in the handling of your solid waste up to and including its disposal. Determine the distance your waste travels from its source to its point of disposal.

Solid Waste Inventory
Data Sheet

Name _____

Section _____

Paper	Glass	Metal	Plastic
1 shopping bag	3 soda pop bottles	2 beverage cans	1 milk jug
2 paper cups			

1.   Were you surprised by the amount of solid waste you generated? Explain your answer.

2.   In what ways do you think you could reduce the amount of waste you generate? Are you willing to do so? Why or why not?

3.   Is there a recycling program in your community? If so, what type(s) of material does it accept?

4.   Of the solid waste you generated during your exercise, what percentage of the total could be recycled in your community?

5.   Why does the volume of solid waste generated in North America continue to grow?

6.   Why do North Americans dispose of more solid waste than Europeans or Asians?

Waste Reduction and Pollution Prevention

Objectives

1. Consider alternative methods of waste management to traditional waste-management strategies such as landfilling and incineration.
2. Identify opportunities to reduce waste generation in a hypothetical production plant and a hypothetical consumer.

Introduction

Solid waste is one of the key environmental challenges facing the United States during the next ten years. While the rate of solid waste generation has been increasing since WWII, the amount of landfill space available to dispose of waste has been steadily decreasing. The demand for available landfill space has increased dramatically during the past several years. This demand has driven up the price of disposing of a ton of garbage in some areas of the country by several hundred percent. As the price of land disposal increases, alternative methods of waste management will become competitive.

One method of solid waste management that is already competitive with land disposal in many instances is pollution prevention. Pollution prevention can also be thought of as waste prevention. Pollution prevention includes any change in process or materials that diverts matter from the waste stream generated by a manufacturing plant, a household, or any human activity. With limits to conventional waste-management practices becoming clearly evident, an ounce of pollution prevention is truly worth a pound of cure.

Most commonly, pollution prevention is thought of as the three Rs: Reduce, Reuse, Recycle. Pollution prevention can also include other approaches including activities such as:

- Improving product durability so that products last longer and therefore need to be replaced less often.

- Improving product updatability. By making a product updatable, a manufacturer can increase a product's life cycle. Such strategies can decrease the amount of waste generated because products are updated rather than replaced.

Indeed, pollution prevention can be thought of as a way not only to reduce waste generation but also to reduce resource consumption.

Procedures

Your instructor will have asked you in advance to sort a can full of garbage at home to separate recyclable from nonrecyclable materials and to note what was nonrecyclable. You can use the table following this paragraph to record the materials you separate. As a class you will travel to the nearest recycling center. This trip will familiarize you with the location of the center and the procedures for preparing material for drop-off (for example, tieing newspapers into bundles and crushing cans).

Type of waste	Estimated quantity	Recyclable	Compostable
Paper products			
Yard waste			
Food waste from plant sources			
Food waste from animal sources			
Metals			
Glass			
Plastics			
Rubber			
Wood			
Other			

One form of waste reduction that you might not have seen at a recycling center is waste composting. Most organic wastes from household and industry can be combined with soils and processed as compost. Composting involves mixing organic wastes with soil. When maintained at proper temperature and moisture levels, this mixture is decomposed by aerobic bacteria to form a rich, dark-brown humus loaded with nutrients and organic matter.

Once you get back to the lab, you will begin the second half of this exercise, building and assembling a compost bin. Your instructor will provide you with the materials to begin this process:

1. Four, 4″ × 4″ posts
2. 18′ wire fencing
3. Necessary tools to sink posts and attach wire
4. Enough soil to form two or three layers, each 1″–2″ thick and each covering about 5 square feet
5. 4 cups of 10–10–10 fertilizer
6. 2 cups of agricultural lime
7. Enough 6″–8″ twigs to form one layer 5 feet square
8. Organic material to compost (collected from home during the garbage separation exercise) including paper, leaves, grass clippings, kitchen waste, etc. Do not include animal-derived materials, as they will accentuate odors and attract animals.

Once the bin has been constructed, add the compost resources as follows:

1. Begin with a layer of twigs to provide aeration.
2. Mix soil, fertilizer and lime and stratify with alternating layers of organic material (note organic material should start just above twigs and soil mix should be on top of heap).

Once the bin has been constructed and material has begun to be placed inside, it is important to maintain moisture and oxygen levels within certain ranges. If you notice strong odors, try increasing the oxygen available to the heap. If this does not work, add sawdust to the pile. Proper moisture levels can be achieved by making a depression in the top of the heap to collect rainwater. The heap must be turned or rotated on a monthly basis. During winter months, cover the bin with a tarp.

Your instructor will divide the class into groups. Each group will be responsible, throughout the semester, for providing organic material to the pile, maintaining the bin, rotating the material, and recording the appearance of the material. At the end of the semester, the class will compile and analyze the data collected throughout the term. After you finish the exercise, be sure to compost the result of this analysis!

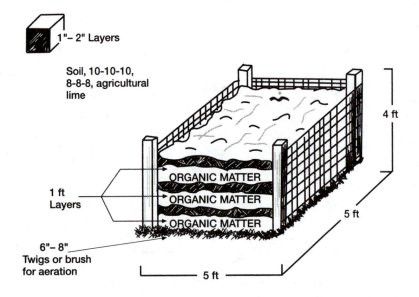

Waste Reduction and Pollution Prevention Data Sheet

Name _____

Section _____

1. As indicated by the data collected when separating your garbage, what contribution to waste reduction can composting make in your household?

2. Of the nonrecyclable and noncompostable materials in your household waste stream, what materials could be altered to make them recyclable or compostable?

3. Some materials have no possibility for recycling, reuse, or composting. What can be done to prevent these wastes from reaching your wastebasket? What types of product design changes could help? What types of product use practices could help?

4. What uses can you suggest for the compost you have created? What local uses are likely? What activities have the potential to consume large quantities of compost? Are any of these activities located in your geographic area?

Source Reduction

Objectives

1. Understand source reduction, pollution prevention and waste minimization.
2. Become aware of, and understand opportunities for practicing source reduction.

Introduction

Many types of pollution and waste take gaseous or liquid forms and not solid forms. Included in these categories are stack emissions from coal power plants, tail pipe emissions from cars, trucks, and diesel locomotives, liquid wastes from chemical manufacturing facilities and more. In many cases, liquid and gaseous wastes are treated to reduce environmentally harmful constituents in the waste and are then released into the environment.

Liquid and gaseous wastes are different from solid wastes in that they are much more difficult to contain, transport and manage once they have been generated. Often the best strategy for keeping such wastes from reaching the environment is to employ source reduction strategies to reduce or eliminate the use of the materials that lead to the generation of the waste. Examples of source reduction include:

- Substituting safer chemicals in processes and products
- Reformulating products
- Improving operations and maintenance procedures
- Practicing closed-loop recycling of processing materials
- Making processes more energy and input efficient
- Increasing product operational life span

Applying source reduction strategies in manufacturing and product design involves planning, research and commitment. A good example of source reduction strategies in product design and chemical substitution is the reduced use of the chlorofluorocarbon Freon-12 (CCl_2F_2), the refrigerant used in automotive air-conditioners and household freezers, refrigerators, and air-conditioners.

Source reduction in the case of Freon-12 has involved an international effort to find environmentally less-destructive substitutes for the chemical and to redesign Freon using products to be both compatible with the new refrigerants and more serviceable.

The development of Freon replacing substitutes has involved efforts of many manufacturing companies, both those involved in making products that incorporate refrigerants in their design, and companies that produce refrigerants for use in consumer products. The outcome of this effort has been new, less-environmentally damaging chemicals that can be substituted for Freon-12.

Another problem to overcome involved the servicing of Freon-12 using products. Automotive air-conditioners utilize a closed-loop compressor system to circulate refrigerant and cool the inside of the car. However, until recently many automotive air-conditioning systems were not designed to allow refrigerant to be collected from the system when being serviced. Thus, if a car was taken into the shop for a poorly performing air-conditioning unit, the resulting repair often involved venting the refrigerant to the atmosphere. This venting would cause 500–1,000 grams of Freon-12 to be released to the atmosphere in every such instance.

The solution to this problem involved the design and installation of valves to allow refrigerant to be removed from and added to the system without venting the refrigerant to the atmosphere. Several companies have been involved in this design work, and successful "quick-connect" valves have been designed and are now being used on many vehicle air-conditioning systems.

Procedure

In the case of Freon-12, the source reduction strategies employed included: 1) finding a less-environmentally damaging substitute and 2) redesigning the product to eliminate venting the material to the environment. Consider the following product design and manufacturing example and work with your classmates to devise possible source reduction strategies for the business.

Consider the case of a small electroplating company that uses as its primary inputs, chemical solvents, metals, electricity, and labor. The manufacturing process the company uses includes a series of tanks where products are dipped for predetermined intervals, and then removed, dried, and taken to another tank to be dipped. After the solution in a tank is used a certain amount of time, it becomes unusable because of suspended solids or other impurities. These solutions are removed and the tanks are cleaned and filled with new chemicals. The spent solution is put in storage for pick up by a waste management company with which the firm works on a contract basis.

The company keeps its production area well ventilated to avoid exposing its workers to elevated levels of harmful airborne chemicals, and all workers wear protective clothing to avoid direct contact with harmful liquids and solids. The primary waste streams created by the company include:

- Spent plating solutions,
- Plating sludges,
- Cleaning wastes,
- Waste rinse water, and
- Treatment liquids.

Source Reduction
Data Sheet

1. Identify opportunities in the scenario to reduce waste at the source. Refer to the source reduction strategies for ideas. When thinking of source reduction, in addition to thinking about the strategies already listed, think of strategies for each of the forms waste can take: solid, liquid, and gaseous.

2. Do you think the company could save money by implementing source reduction strategies in its manufacturing process? If so, explain how savings might be achieved and show possible opportunities for savings by listing sources of potential savings.

3. Source reduction takes planning and commitment. What are the barriers the company faces when implementing a sources reduction strategy and what kinds of steps do you think the managers of the firm should undertake to ensure that the source reduction strategies are implemented?

4. This lab exercise has focused on the changes manufacturing firms can make in the design and manufacture of consumer products. What changes can be made by the public and by individuals such as you to promote source reduction behavior? Give examples.

Campus Environmental Inventory

Objectives

1. Show that the solutions to many environmental problems are within reach of the individual.
2. Provide a framework by which the environmental concerns and impacts on your campus can be evaluated.
3. Develop actions and policies on your campus to reduce environmental impact.

Safety Considerations: No safety considerations are necessary for this exercise.

Introduction

Because of the diversity of campus activities, colleges and universities represent a microcosm of the environmental issues that confront the world. In addition many campuses are large enough to have a sizable impact on the environment, both through their consumption of large quantities of energy and resources and their generation of large amounts of waste.

Students can work toward making campuses positive examples for environmentally sound management and development. The purpose of this exercise is to provide a framework for researching environmental issues on your campus. Seven environmental issues are included in this exercise. Each issue provides an outline of the data needed and potential sources of information. Although each campus has a unique set of environmental problems and opportunities for change, this exercise raises issues that are common to many campuses and suggests general strategies for resolving environmental conflicts. This exercise does not include all of the potential environmental issues facing campuses. You may wish to address other issues such as open space and wildlife, construction, medical waste, or wastewater and storm runoff. You can use the format of this exercise to investigate other issues.

Procedures

1. Read through the exercise.
2. Using the suggested steps become familiar with your campus.
3. Complete your campus environmental audit response form.
4. Develop recommendations.

Becoming Familiar with Your Campus

Before you begin your research into specific environmental issues, it is important to familiarize yourself with the campus facilities and administrative structure. The following list suggests the types of people who should be contacted and the types of information that should be collected.

1. *Review a map of your campus and area.* Develop a sense of place with regard to your campus.
2. *Review campus and local community newspapers.* Does your university have a history of providing leadership? Does it influence policies at various governmental levels?
3. *Obtain administrative organizational charts.* Who holds the authority to make changes? Which decision-making body sets environmental policies on campus?
4. *Contact your student association.* Find out about campus unions and local environmental and community groups.
5. *Review the range of facilities on your campus.* Include food services, maintenance, utilities, housing, and research facilities. Who makes decisions regarding each facility?

Where to Obtain Information

In this exercise there are questions directed at solid waste, hazardous waste, pesticides, water consumption, energy consumption, transportation, and purchasing. As previously mentioned you may wish to expand this list to include other environmental concerns. The following is a list of where you can find some of the information to complete your campus environmental inventory.

1.   Contact the facilities maintenance department for information about garbage volumes, costs, collection processes, and disposal costs.
2.   Talk with representatives from the company that hauls the waste from your campus. Custodial staff may also be a source of information.
3.   Food service managers can provide information on the use of plastic, paper, and other disposable material.
4.   Contact local recycling centers and your local Public Works Department for information.
5.   Talk with professors and students in departments that handle hazardous substances and ask about their procedures for waste disposal.
6.   The campus purchasing office may also have information on the amount of pesticides ordered and the cost of pest control.
7.   Contact your campus maintenance department for information regarding water use and water monitoring.
8.   Contact campus physical plant personnel or the energy manager for information about campus energy consumption and conservation.
9.   Be creative in obtaining information. Approach this exercise as a detective would approach a case. Go after the needed information.

Campus Environmental Inventory
Data Sheet

Name _____

Section _____

Solid Waste

1. How much total solid waste does your campus generate annually?

2. Does your campus have a recycling program?

3. What were the costs of solid waste disposal last year? What was the cost of the recycling program last year?

4. Describe any programs your campus has implemented to encourage source reduction.

Hazardous Waste

1. How much hazardous waste does your campus generate annually?

2. How is the hazardous waste disposed of?

3. Has your campus initiated a hazardous waste reduction program? If so, describe it, including date of implementation and cost savings to date.

Pesticides

1. What are the most commonly used pesticides on your campus?

2. What is the total volume of pesticides used on your campus?

3. Has your campus initiated an integrated pest management program? If so, describe it.

Transportation

1. How many vehicles travel onto campus daily?

2. What percentage of your campus area is devoted to roads and parking lots?

3. Has your campus initiated a program to encourage alternatives to the single-occupant vehicle? Ride-sharing programs? Bike lanes? Public transportation? If so, describe them and include implementation dates.

Water Consumption

1.   How much water did your campus consume last year (in liters)?

2.   Describe any water conservation programs on your campus; include the date of implementation.

3.   What were the water utility costs for the campus last year? Are water utility costs for the campus increasing or decreasing? Why?

Energy Consumption

1.   How much energy did your campus consume last year, and what were the costs associated with each type of fuel?

     _____ KWH electricity

     _____ BTU natural gas

     _____ BTU fuel oil

     _____ Other:

     _____ Total

2.   Does your campus have an energy conservation program? If so, describe it; include date of implementation and cost savings to date.

3.   Does your campus have a plan to encourage the use of renewable energy sources? If so, describe it; include date of implementation and cost savings to date.

Purchasing

1.   Does your campus buy recycled paper?

2.   What programs and policies have been established on your campus to promote the use of ecologically sound products (for example, organic produce, compact fluorescent lightbulbs)?

Probability and Risk

Objectives

1. Develop an understanding of the role of probability in assessing risk.
2. Understand the importance of cumulative factors in assessing risk.
3. Recognize that we all engage in behaviors that involve risk.
4. Recognize that assessing risk does not prevent random events that could be harmful.

Safety Considerations: There are no specific safety considerations for this exercise.

Introduction

Risk assessment is a difficult exercise. Some risks are easy to quantify. For example, if one looks at large numbers of individuals who snow ski, it is possible to state the incident rate for broken bones. Likewise intravenous drug users and those who practice unsafe sex have a substantially greater risk of getting AIDS. Other events are random and less predictable. For example, people killed in commercial plane crashes have very little control over their fate. Similarly there is little those who are killed or injured because of natural disasters can do to reduce their risk.

Procedures

1. Complete the personal risk inventory and assign "risk points."
2. Use playing cards to simulate random events.

Risk Determination

This exercise will assign risk points to some of your personal characteristics and also simulate random events. Complete the following chart and assign yourself risk points.

Male	2 risk points
Female	1 risk point
If you smoke	2 risk points
If you do not smoke	0 risk points
Do you use a seat belt?	1 risk point
If you do not use a seat belt	2 risk points
Are you considered overweight?	1 risk point
If you drive 0 kilometers per week	0 risk points
If you drive 1–25 kilometers per week	1 risk point
If you drive 25–100 kilometers per week	2 risk points

Your total risk points: _____

Use a deck of playing cards to determine randomly who should be assigned additional risk points. Shuffle the deck and deal everyone in the class one card. All those who receive an ace receive an additional risk point. Deal twenty rounds. As soon as you accumulate ten risk points you are "dead" and do not participate in the rest of the exercise. Your instructor may wish to have you repeat the exercise to see if the results are similar over two or three repeats of the exercise.

Probability and Risk
Data Sheet

Name _____

Section _____

1. At the end of twenty rounds how many people in the class were still "alive"?

2. What were the highest, lowest, and average number of risk points that people began with?

   Highest

   Lowest

   Average

3. Were people who were in higher-risk categories more likely to be "dead" at the end of the exercise? Are all people subject to the same risk?

4. Were initial risky behaviors or random events more important in deciding who remained alive at the end of the exercise?

5. List five personal behaviors that could increase your risk of death or injury.

6. List five environmental factors that could increase your risk of death or injury.

Economics of Energy Consumption

Objectives

1. Become familiar with the energy consumption information available to consumers.
2. Calculate the energy consumption and compare the cost of energy consumption to the initial purchase price.
3. Evaluate the long-term cost of purchases of energy-consuming products.

Safety Considerations: There are no specific safety considerations associated with this exercise.

Introduction

Many of the items we use every day consume energy. The way in which the item is designed influences the amount of energy used. Often items that are inexpensive to purchase may have large costs for the energy to operate them. For example, a poorly insulated refrigerator may be less expensive to manufacture but since it will require more energy to operate may cost more in energy expenses than would a more expensive, better insulated model.

During this exercise you will collect information on a variety of energy-consuming consumer goods and evaluate their energy consumption over a ten-year period. You will need to visit or call appliance and automobile dealers to obtain the information present on the energy stickers required on all such goods.

To obtain the ten-year energy cost assume that the cost of electricity, natural gas, or gasoline will be constant for the next ten years. You can obtain current energy costs by checking your utility bills and local gasoline stations.

Stove, Water Heater, Refrigerator

Obtain information for two different appliances in each category. Make sure the appliances have the same basic features but that there is a considerable difference in price. Avoid models that have lots of gadgets. Go for the utilitarian models. For example, make sure that the refrigerators you compare have the same capacity and do not have ice makers, drinking water dispensers, or other special features.

Lightbulbs

A 100-watt incandescent lightbulb gives about as much light as a 40-watt fluorescent lightbulb, so they are comparable in terms of the amount of light produced. Assume that you will use the lightbulb for 1,000 hours per year and that you will need to replace the incandescent lightbulb each year. (You will need to buy ten incandescent lightbulbs but only one fluorescent.) Calculate the number of kilowatt-hours used by multiplying the wattage of the light by the number of hours used in ten years and dividing that number by 1,000.

Automobiles

Select two automobiles in the same size category with essentially the same equipment. Use the sticker price as the price of the automobile. Assume that you will drive the automobile 20,000 kilometers per year and that 10,000 of those kilometers will be in city traffic and the remaining 10,000 will be highway driving. Assume that the car will last ten years. Do not try to take maintenance costs into account. Assume that fuel prices will be constant over the ten years.

Other

Your instructor may want to assign additional items for you to research.

Economics of Energy Consumption
Data Sheet

Name _____

Section _____

Item	Purchase price	Energy sticker information	Annual energy used	Annual cost of energy used	Ten-year cost (purchase plus ten years of energy)
Gas stove 1					
Gas stove 2					
Electric stove 1					
Electric stove 2					
Refrigerator 1					
Refrigerator 2					
100-watt incandescent light bulb					
40-watt fluorescent light bulb					
Automobile 1					
Automobile 2					
Gas water heater 1					
Gas water heater 2					
Electric water heater 1					
Electric water heater 2					
Other					

1.   At current energy prices, which are least expensive to operate, gas or electric appliances? Is the difference significant?

2.   If you had an old, poorly insulated refrigerator that used twice as much electricity as the best model you researched and you replaced it with the more efficient model, how many years would it take for you to recover the cost of the new refrigerator? Calculate this by dividing the purchase price of the new refrigerator by the number of dollars saved per year because of lower total energy costs.

3.   Determine the price of a central ceiling 40-watt fluorescent light fixture for your room. How many years would it take for you to save enough money on energy costs to pay for the fixture?

4.   What percentage of the purchase price of an automobile is the cost of fuel for one year if you were to drive 20,000 kilometers?

Sustainable Agriculture

Objectives

1. Identify the criteria that must be met to achieve sustainable agriculture.
2. Determine possible short- and long-term environmental, economic, and agricultural policies that could help move the nation and the world toward sustainable agriculture.

Introduction

Sustainable agriculture is one requirement for the future if worldwide food shortages and the associated dislocation of overstressed societies are to be avoided. Current industrialized agricultural practices in the United States often involve the use of capital-intensive equipment to manage large farms. This capital, in the form of tractors, combines, grain silos, and irrigation machinery, often involves frequent tilling of the soil to make the conditions most favorable for a single crop such as corn or soybeans. These farms generally use significant levels of synthetic herbicides, insecticides and fertilizers as well.

Soil erosion is one problem of most agricultural systems. In particular, soil erosion can be very prominent on sloped land or land where erosion barriers are not used. Soil erosion for the capital-intensive farming practices described in the preceding paragraph is also significant because of the constant tilling of the soil.

Beyond erosion, however, less obvious changes are also taking place in the soils, climate, and genetic resources of the world. Overcultivation is reducing the levels of nitrogen, phosphorus, and potassium. In irrigated areas, soil salinization is a serious problem and, in other areas, soil irrigated for decades becomes waterlogged and unable to support crop growth. Climatic changes may cause significant changes in the location of fertile fields in the future and loss of genetic diversity may limit the ability of genetic engineers to combat climate changes, insects, and other sources of crop stress.

Procedure

For any system, whether an economic system, a mechanical system, or a biological system, the concept of sustainability is a concern, and the dynamics of sustainability are surprisingly similar across system types.

1. How do you define sustainable agriculture? What requirements do you think must be met if agriculture is to be sustainable?

2. Does sustainability mean that more or less food will be available under sustainable agricultural systems?

3. Does sustainability in the agricultural sector have ripple effects to other sectors of society? If sustainable agriculture means using locally grown foods, what will this mean for transportation industries? Fuel industries? Agricultural equipment industries? Will agricultural sustainability lead to overall economic sustainability?

4. For the past twenty years, economists have been discussing the loss of natural resource capital such as destruction of forests, degradation of surface waters, and consumption of stock resources such as coal and oil. Some economists suggest that these losses should be subtracted from GNP totals to yield a net GNP figure that reflects environmental resource losses. Under such a system, do you think economic sustainability would be easier to track? Could a smaller but similar system for resource degradation be used for agricultural systems? What types of things would you attempt to measure in such a system?

5. The information provided in the introduction gives you an idea of some of the challenges facing the future of agriculture. If you were a congressperson working in Washington, D.C., what types of policies would you support for agriculture?

**Environmental Economics—
Business and Environment**

Objectives

1. Understand some of the important economic issues associated with the environment.
2. Understand the role of economics in environmental decision making and environmental policy.
3. Understand why firms have economic incentives to pollute.

Introduction

Economic theories are used to analyze how people organize themselves to produce, consume, and trade goods and services. Economics looks at market outcomes and people's preferences for various goods and services and seeks to identify efficient ways to accomplish goals.

One of the basic theories of environmental economics is the theory of cost internalization and externalization. Internalized costs are the costs of production and consumption that are reflected in the marketplace. For instance, suppose company X produces pizza. The cost of the pizza reflects the expenses the pizza company incurred while producing the pizza (plus profit) including items such as:

1. Pizza sauce, cheese, and dough
2. Pizza toppings
3. Pizza box
4. Pizza delivery costs (car, driver, gas, car maintenance, etc.)
5. Store costs (rent, heat, lights, water, taxes)
6. Legal fees
7. Advertising
8. Any other items appearing on the company's expense sheet

The cost of the pizza to pizza customer Y is the cost of the pizza itself, any tip given to the driver, the cost of the phone call (if any), and the cost of garbage disposal (if any).

All the costs listed above for company X and consumer Y are reflected in the market exchange of the pizza. For example, should the price of pizza boxes increase, the pizza company must either raise the price of pizza, find an alternative to pizza boxes, or reduce profits. Thus pizza box expenses are internal to the market.

Not all the costs borne by society, however, are reflected in the price of the pizza. Who pays for the health losses created by the pizza delivery car polluting the air? Who pays for the waste management needed when pizza boxes are thrown away? Who pays for the watershed degradation such as stream-bank erosion and stream eutrophication caused by dairy cattle whose milk is used to make pizza cheese? Who pays for all these costs that are *external* to the pizza market and therefore not reflected in the price of pizza? The answer: We all do!

If we all share these costs equally, then it might be considered OK to pollute. If the problem is created collectively, it can be resolved by collective action.

Well . . . as usual it's not that simple. In looking at the hypothetical health costs for this example, we see that health problems caused by polluting pizza delivery vehicles may be borne by someone entirely removed from the pizza market. The pizza company has no incentive to reduce this pollution so it maintains its cars by expending the least dollars per vehicle to keep them functionally and legally roadworthy. Yet, the people who become ill from elevated air pollution incur costs when getting treated. These costs will be paid for by the sick individual, the individual's health insurance company, or by medicare (federal health insurance system). These costs are external to the pizza company's expenses and to the cost of the pizza to the consumer. Hence, they are called market externalities.

As this example illustrates, the market price for pizza does not reflect the true costs to society of producing and consuming pizza.

Procedures

1. List three types of market externalities you can identify for a manufacturing business. List three types of market externalities you can identify for your household.

2. Common tools employed by policy makers to force companies and individuals to "internalize" the costs of production and consumption are:

    - Developing regulations and fines
    - Granting tradable pollution rights
    - Enacting green taxes
    - Charging user fees

    Could any of these policy instruments be used in the pizza example on the preceding page? Explain.

3. Government-subsidized activities such as low-cost western grazing permits, timber permits, mining permits, and other subsidized activities can increase the rate of resource extraction above rates determined by market equilibrium forces. How do subsidy policies such as these compare with the policies mentioned in question 2? Would they tend to lead to greater environmental degradation? Are they consistent with environmental initiatives?

4. For further information on the concept of market externalities, refer to Garret Hardin's 1968 essay "The Tragedy of the Commons." This essay has had impacts on economics, political science, and on the management of renewable resources.

Objectives

1. Become familiar with concepts of environmental risk and risk assessment.
2. Become familiar with the process of risk-based policy development.
3. Work through a hypothetical risk-based policy decision.

Introduction

We've all heard how certain sports like skydiving or motorcycle racing are "risky." One wrong move and a motorcycle racer could be dead or severely injured. In environmental science, risk is the probability of suffering harm from some type of danger. The higher the risk associated with an activity or situation, the more likely it is that harm will result.

Common hazards people confront daily include:

- Cultural hazards (including motorcycle racing for those of us that race motorcycles)

- Pollution hazards (from human-made toxins in the air, food, and water)

- Physical hazards (act-of-God events such as floods, earthquakes and hurricanes)

- Biological hazards (viruses, parasites, predators in natural ecosystems)

Risk assessment is used to determine the risk posed to human health or the environment that may result from certain exposures. Risk assessment involves using data, assumptions, and modeling to estimate the probability of harm. The Science Advisory Board for the U.S. Environmental Protection Agency has recommended that environmental policy decisions be risk-based.

Risk assessments often determine the risk to a population from a particular pollutant or waste stream by combining two factors:

1. The number of people likely to be exposed to the pollution
2. The chance (risk) of illness, death or harm likely to come to each person exposed

Hypothetical Risk Assessment Case

Imagine you are a government official in a position of authority in the federal government. You are faced with the following information.

Example 1

A previously unknown harmful effect from air-cushioned tennis shoes (in our example everybody has worn these shoes at least once) was just discovered. Anyone who has ever worn such a shoe is equally at risk. This harmful effect will start to become evident among a small percentage of the population. It is estimated that 600 deaths will result each year if no action is taken.

Another previously unknown harmful health effect from standing with your back to the wind (in our example everybody does this) is discovered the same day. This harmful effect will become evident in a small percentage of the population over the next year as well. It is estimated that 600 deaths will result each year if no action is taken.

There are two vaccination programs you can implement but you have limited resources and must develop one vaccine program first and then the other.

In the above example, which vaccine program would you develop first—the shoe vaccine or the wind vaccine? What factors would you use to make your decision? What additional information would be helpful to you when making a decision?

Example 2

Now the situation is the same as above but additional information is available about each of the vaccines. For both vaccines, the cost and time, six months in each case, necessary to develop the vaccine is the same. And in each case, owing to limited laboratory space nationally, one vaccine must be developed and then the other. In addition, you have been informed that one year from now, 100%-effective vaccines will be introduced for both the shoe and the wind effect.

The vaccine for the shoe effect will be successful in preventing death in 33% of all shoe-related illnesses.

The vaccine for the wind effect has a 66.6% chance that it will be completely ineffective, resulting in death for all wind-related illnesses.

Procedures

1. In the preceding examples, which vaccine would you choose to pursue first? Why? Are the vaccination programs really different? If so, how?

2. How would your decision be affected if one vaccine took longer to develop?

3. How would your decision be affected if the time frame for development were the same for each vaccine but the costs of the two vaccination programs were vastly different?

4. One yardstick for environmental regulations is often the number of cancer deaths reduced per dollar needed to implement the regulation. Thus, if a new environmental regulation costs $100 million to implement by industry and government and results in ten fewer cancer-related deaths, the cost effectiveness of the regulation is $10 million/cancer-related deaths. Can you think of an example of an environmental regulation that is either cost-effective or cost-ineffective using this yardstick?

5. Using the cost-effectiveness criteria outlined in question 4, would you guess that the U.S. Clean Air Act or the U.S. Resource Conservation and Recovery Act (this act regulates hazardous waste generation, transportation, and disposal) is more effective at reducing risk? What is the basis of your choice? (Hint: Because air is a free-flowing medium, more people are exposed to air pollution than to pollution resulting from contamination of groundwater and soils caused by improper hazardous waste management.)

6. What goal should legislators and regulators strive for in reducing risk in the legislation and regulations they write?

7. The Science Advisory Board for the U.S. Environmental Protection Agency apparently believes in value or risk assessment as a critical factor in the development of environmental regulations. Besides risk and implementation cost, what are some other criteria that may be used to assess an environmental regulation?

Objectives

1. Develop a working understanding of some critical *concepts* and *terms* routinely used in health and ecological risk assessments, risk management decisions, and public risk communications.
2. Rank your perception of the relative risk associated with several potential concerns to a desirable quality of health and environment.

Introduction

How are the people living near a source of environmental pollution, such as a gas station, power plant, trash-to-energy incinerator, contaminated property, or landfill, supposed to have confidence in their safety? Similarly the food we eat may have a detectable level of a pesticide contaminant, the water we drink may contain chemicals that cause cancer in laboratory test animals, or the fish we catch may be contaminated by past pollution. Historically these difficult questions were addressed on a case-by-case basis through government-industry negotiations. Today the public is receiving more specific regulation of the risks associated with significant environmental challenges and more objectivity in the decision making. This drive for more objectivity has forced the immature science of estimating health risk into the forefront of environmental policy.

The EPA recognizes that current risk assessment methodologies significantly overestimate actual risk by factors from 10,000 to 30,000, and the true risk for a given situation may be zero. Risk managers are supposed to take the conservatism of a calculated risk value into account when making environmental decisions, but the role of risk assessment and risk management have become so blurred that public policy now makes it sounds like "one-in-a-million" or "one-in-a-hundred-thousand" are acceptable levels of risk. Does anybody really know what these levels of calculated risk mean?

Contrary to what it sounds like, one-in-a-million does *not* mean that, for each million exposed individuals, one individual will (or might or is expected to) get cancer. What it does mean is a regulatory benchmark for a standardized calculation methodology. These "acceptable levels of calculated risk" are actually relative risks and are useful when comparing one regulatory decision with another or when setting priorities for environmental protection programs.

From a public communication perspective, discussions about calculated risk levels usually do more harm than good. They outrage those bearing the risk and polarize the decision makers. For example, in the 1950s there were few environmental health issues associated with the siting of a gas station. Today the neighbors of gas stations are concerned about the cancer risk associated with the benzene fumes from gasoline when we fill our cars, and they are concerned about the gasoline that contaminates the property from leaking underground storage tanks. Can the risk associated with living near a gas station be quantified?

The field of risk assessment is new. It is also a field that has few, if any, absolutes. Perception of risk and true risk are not the same. The question remains, however; how do you separate perception from fact?

Procedures

1. Read the exercise.
2. Select two or three sets of terms in table 1 and write a brief paragraph describing what you think each term in the set means, how the terms are similar or different, and how each term might be used to describe environmental risk.
3. Select ten of the challenges to a desirable environmental quality and public health listed in Exercise 41.2 and then rank them on the work sheet with the fraction of 100 units of environmental protection resources that you feel should be allocated to each challenge.

Laboratory Exercises

Exercise 41.1 Working Knowledge of Risk Terms and Concepts

Your instructor will assign you two or three sets of items in table 1 to research and report on. Write a brief paragraph for each set of terms that compares and contrasts the concepts in the space provided on the data sheet. Then describe in class why these terms are important in assessing and managing risk, using relevant examples whenever possible. Copies of recent news and scientific articles of environmental issues involving risk would be helpful for this exercise. Your views on some of these terms and concepts may become important in future public opinion and policy.

Table 1   Sets of Risk Terms and Concepts.

Set A.	Voluntary versus involuntary risk
Set B.	Health risk versus environmental risk or hazard
Set C.	One-in-a-million versus one-in-a-hundred-thousand
Set D.	Genotoxic versus non-genotoxic carcinogen
Set E.	Actuarial risk versus extrapolated risk
Set F.	Threshold versus non-threshold mechanism
Set G.	Upper bound on risk versus most likely estimate of risk
Set H.	Linearized multistage model versus no observable effect level
Set I.	Carcinogenicity versus mutagenicity or teratogenicity
Set J.	Maximum exposed individual versus actual exposed individual
Set K.	Chemicals classified A, B1, or B2 carcinogens versus C carcinogens
Set L.	Equipment failure risk versus cancer risk
Set M.	Individual risk versus population risk
Set N.	Risk assessment versus risk management

Exercise 41.2 Ranking Environmental Challenges

Each of the following represents a challenge to a desirable environmental quality and to public health and is currently the focus of risk-reduction programs. Place each in order of its need for *additional resources* and offer a brief rationale for your ranking of the top three priorities on the data sheet. Discuss the results in class. Is there a consensus in your class on the rankings? Describe why you ranked as you did.

Residual pesticides in food	Hazardous waste site cleanups
Landfill operations	Contaminated property
Industrial air emissions	Municipal wastewater treatment
Trash-to-energy incinerators	Agricultural runoff
Contaminated sediments	Auto emissions
Chemical accidents	Nuclear power
Industrial wastewater treatment	Other _____

Risk Terms and Concepts Definition

First set

Second set

Third set

Ranking Environmental Risk

Priority of environmental challenges in need of additional resources	Fraction of 100 units of additional resources	Your rationale
1.		
2.		
3.		
4.		
5.		
6.		
7.		
8.		
9.		
10.	100 Units	

Environmental Law and Decision-Making

Objectives

1. Become aware of environmentalists' views of how environmental decision-making should be made.
2. View the array of strategies that may be used in an environmental dispute.
3. Understand the role of environmental law in environmental decision-making.

Introduction

Environmental law is relatively new—first appearing in law school curricula around 1970. Since 1970, environmental law has grown dramatically with environmental law courses being taught at over 150 law schools throughout the country and environmental lawyers making up one of the fastest-growing areas of the legal profession.

The first law of ecology holds that everything is connected to everything else. In environmental decision-making, applying the first law of ecology means taking a broad view of the potential costs and benefits involved in a proposed decision and taking a careful look at the potential alternatives to the proposed course of action. Environmental decision-making is based on a long-term perspective and a broad accounting of costs and benefits.

Law is just one course of action environmentalists employ in environmental disputes. The broad array of strategies that may be employed in an environmental dispute include:

- Marketplace remedies
- Petitioning agencies
- Petitioning legislatures
- The media
- Legal suits

Legal suits can include civil suits, constitutional claims, public trust-based suits, statutorily based actions, administrative law litigation, or international law. As you can see from the list above, legal action is just one of a number of strategies used to influence environmental decision-making.

Procedure

In this lab exercise, two groups in class will be asked to take opposing sides in a mock trial involving a large industrial company that operates an ore-processing facility outside the city limits of a small town. The plant emits dirt, smoke, and vibrations from the mining and processing activities in which it engages. When the wind blows in the right direction, fine dust particles and dirt from the plant's operation noticeably collect on the roofs and eaves of the houses in the adjacent neighborhood. Although the dirt has been tested and found to be nontoxic, its unsightly appearance has reduced property values in the neighborhood.

Separate the class into two groups. One group will represent the local citizens living in the neighborhood adjacent to the plant. The other group will represent the company that owns the plant. Your instructor will act as judge. Three students will be appointed to the jury.

As your group prepares arguments for its side of the case, be sure to consider the following questions:

1. Why does the mining company not voluntarily reduce or eliminate dust emissions? What is the decision-making process the company employs that results in dust emissions?
2. Does it matter who was there first? Company or neighborhood?
3. In addition to the dust on the homes, what other types of damages might the home owners be experiencing?
4. What possible outcomes might the citizens seek against the company?
5. What must the court consider in balancing the arguments of the two sides of this case?
6. How might a court case lead to a broader accounting of the costs and benefits of operating the mining facility and the decisions that affect its operation?

Source: The information for this lab was drawn from *Environmental Law and Policy: Nature, Law and Society* by Plater, Abrahms, and Goldfarb, 1992, West Publishing. The case presented in outline above is analogous to the classic environmental case *Boomer et al.* v. *Atlantic Cement Company,* New York Court of Appeals, 1970 (26 N.Y.2d 219, 257 N.E.2d 870, 309 N.Y.S.2d 312). This lab involves a significant level of instructor preparation.

NGOs—NonGovernmental Organizations

Objectives

1. Become familiar with the concept of NGOs and typical NGO missions.
2. Understand the role of NGOs in environmental decision-making and environmental policy.

Introduction

Nongovernmental organizations, or NGOs, have been organized around virtually every conceivable issue. Some NGOs, such as trade associations, represent the desires of industry (lobby) in state and federal legislatures or coordinate certain types of information collection and distribution for industry. Other NGOs represent consumer rights in the national forum. Yet others advocate policies for improved environmental quality, provide information on environmental careers, or conduct research on environmental issues. You may be familiar with several environmentally oriented local and national NGOs.

Because NGOs with environmental agendas have become so numerous, numbering almost 7,000 in the United States alone, they merit a closer look. Indeed, the environmental movement throughout the world is driven by the actions of thousands of NGOs that have organized around pollution, energy consumption and energy sources, consumerism, population, biodiversity, and other issues.

Procedures

1. Your instructor will lead a class discussion addressing why NGOs exist. You will answer questions such as:

   - Why does society need NGOs?

   - What roles do NGOs play in decision making?

   - What services do NGOs provide the public?

2. Next your instructor will help you list some of the activities of NGOs. On the chalkboard list the following headings and spend twenty minutes filling in the table.

Missions of NGOs (type of work they do)	Environmental Issues Addressed (topics they address)	Examples of NGO Activities

3.   NGOs often form coalitions to work together, pool resources, and improve their overall effectiveness. Starting in 1988, several hundred local and national NGOs in the United States began forming a coalition called the "wise-use movement." The wise-use movement advocates such practices as:

   - Harvesting all old-growth forests and replacing them with tree plantations
   - Eliminating restrictions on wetland development
   - Opening most land in the National Wilderness Preservation System for mineral and energy production, off-road vehicle recreation, and development

Much of the financial support for the wise-use coalition comes from developers, timber producers, ranchers, mining interests, and petroleum producers.

   What effects do you think the wise-use movement will have on environmental quality? How do you think the wise-use movement will affect involvement in environmental NGOs?

Objectives

1. Become aware of the role of environmental science in environmental decision-making.
2. Consider avenues of future environmental research.

Introduction

Back in the 1880s a company and a landowner located several hundred yards apart in New York state were enjoying a good relationship. The company, as part of its operations, regularly released residues and wastes from an oil-refining operation onto the ground. After some time, the landowner, who used groundwater from a well located on site (several hundred yards from the refining operation) began to notice contamination in the well water, which rendered the wells unfit for use.

In an early environmental lawsuit, the landowner sued the refining company for destroying the wells. The court found in favor of the oil company. The court opinion at the time stated: "It is only in exceptional cases that the channels of subterranean streams are known and their courses defined." The court continued, "In the absence of knowledge as to the existence of such subterranean water courses, . . . there can be no liability when such courses become contaminated." The presiding judge in the case described the then-new concept of groundwater flows as "subterranean water courses." That these "courses" might carry wastes several hundred yards (or more!) was not common knowledge even 100 years ago.

Today, hydrogeologists know that water is below the ground surface throughout the world. These sources of groundwater are often called *aquifers*. Hydrogeologists are able not only to identify the level and rate of flow of water in aquifers, but they often can also identify different levels of aquifers in the subsurface rock. These aquifers often have different flow rates and flow directions and this information can be critical in determining the spreading and containment of groundwater contamination. Needless to say, so much is known about groundwater flows that an extensive framework of federal and state laws exist to protect the quality of our groundwater resources. Portions of these laws are, to a substantial extent, based on the work of hydrogeologists.

In the example above, the work of geologists and hydrogeologists in the years since the late 1800s has established links between the rate and flow of groundwater and the spread of in-ground pollution. These links have allowed better environmental decision-making and protection. Hydrogeology is but one example of the role science can play in environmental decision-making and environmental protection. By making links between environmental causes and environmental effects, scientists are continually providing environmental decision-makers and those involved in environmental protection and cleanup with better tools for addressing environmental problems.

Procedure

1. Choose an environmental dispute from one of the following suggestions or choose your own environmental issue by class vote. Possible environmental disputes include:

    *   The energy debate presented in Exercise 23

    *   The spotted owl controversy in the Pacific Northwest

    *   Water pollution on the Great Lakes, or pesticide pollution on the Mississippi River

    *   Air pollution in urban–industrial areas

    *   The use of nuclear energy for electricity generation

    *   The use of coal for electricity generation

2. Once you have chosen a topic, make a list, with your classmates, that shows the types of arguments that might be made for and against various solutions to your issue.

3. Once you have listed several arguments, categorize each as mainly:
    A. Ethical or moral statements
    B. Scientific statements
    C. Economic issues
    D. Public opinion
    E. Political issues

4. Your list developed in step 2 and categorized in step 3 now shows some possible areas where science might be able to contribute to the resolution of an environmental issue or dispute. What types of scientific inquiry might help resolve the arguments listed in your particular controversy?

5. Do you see any similarities between your answer to step 4 and the scenario for hydrogeology outlined in the example at the beginning of this lab exercise?

6. What types of environmental research do you feel are prominent today? What types of atmospheric environmental research are prominent? What types of biological environmental research are prominent? What types of geologic and oceanic research are prominent?

Objectives

1. Understand environmental justice.
2. Become aware of, and understand issues associated with, the environmental justice movement.
3. Suggest changes that will need to occur to resolve environmental justice concerns.

Introduction

During the past decade, university researchers have established links between the location of hazardous waste disposal, storage, and treatment facilities and economically deprived communities. These communities often have predominantly minority populations. One prominent study found that the siting of hazardous waste facilities was highly correlated with African-American communities. In other words, African-American communities are most likely to have hazardous waste facilities located within them.

These findings fueled the "Environmental Justice Movement," which linked social issues with the broad diversity of environmental issues—not just hazardous-waste facility siting. Other terms used for environmental justice include *environmental equity* and *environmental racism.*

The environmental justice movement is a grassroots movement composed of thousands of local groups, working at the local level to achieve environmental justice for all.

Procedures

1. List five societal forces you think contribute to environmental injustices. Why are certain communities more likely to be hosts for hazardous waste facilities?

2. Your instructor will lead a class discussion addressing the environmental justice movement, its history, its approach, and its role in society. You will answer questions such as:

    *   What role does the environmental justice movement play in society?

    *   What roles do environmental justice organizations play in decision making?

    *   What services do environmental justice organizations provide the public?

3. Is environmental justice an efficient approach to environmental policy? Good environmental policy achieves the most reduction in environmental risk for the least cost. In some sense, the environmental justice movement, by providing resources to environmentally degraded urban areas (high environmental risk × high population density = high risk to human health), could provide large improvements for little cost.

    What aspects of the environmental justice movement may be thought of as inefficient? Is efficiency a relevant issue to evaluate the contribution of the environmental justice movement to society?

4. Environmental justice is not limited to environmental issues within one country. International shipments of waste from developed countries to less-developed countries can be another form of environmental injustice. Are you aware of any international environmental justice issues or events? What types of solutions to international environmental justice would you propose?

Environmental Ethics

Objectives

1. Become aware of ethical arguments for environmental protection.
2. Consider an ethical perspective toward environmental issues.

Introduction

When presented with an environmental protection decision or issue in a legislature, in the courts, or in a local government, the question that often arises is "Why should we protect this particular aspect of the environment—this lake, or this river, or this stand of trees?" To this question, environmentalists usually reply with arguments that clearly show the usefulness of protection in terms of lives saved, dollars saved, or costs avoided. In other words they attempt to show a measurable value provided by the protective action.

This approach has sparked a decades-old debate among environmentalists. Many people argue that the environment is more than just a resource for human purposes and for human use and that the natural world has value that cannot be measured in monetary terms. Many people believe that there should be a moral, ethical standard extending beyond the narrow economic and comfort needs of humans.

Procedure

1. Split the class into four groups

   - Mountain
   - River
   - Bear
   - Giant sequoia

2. In each group, look at the world from the perspective of a mountain, a river, a bear, or a giant sequoia and consider the following issues:

   - What is it like to be a mountain, a river, a bear, or a giant sequoia?
   - How old are you?
   - What changes have you observed since the beginning of your existence? Species changes? Climate changes? Topographical changes?
   - What does the future hold for you?
   - What changes would you make in the world if you could?
   - What is more important to you?

     —Growth or stability?

     —Diversity or uniformity?

3. Have the class re-form into a single group and compare the responses to the above questions. How are the responses similar? How are they different?

4. Environmental ethics is one of many analytic filters people use to assess environmental decision-making and environmental issues. Can you think of any environmental laws that take into account the rights or perspectives of other creatures? Are these laws based on economic value or is there an ethical basis? Are you aware of any environmental law cases that have an ethical basis? How would you characterize the spotted owl/lumbering issue in the Pacific Northwest?

Influencing Public Officials

Objectives

1. Become actively involved in an environmental issue or problem.
2. Become informed about an environmental issue or problem and convey your opinion in a letter to an elected official.

Safety Considerations: There are no safety considerations for this exercise.

Introduction

It is important in a democracy that citizens help keep legislators informed. This is especially true with regard to environmental issues. In order to be true representatives of the people, legislators need to know what their constituents think. As a citizen, you can help obtain good legislation on state and national levels by communicating with your elected representatives at the proper time. Too many people never have any contact with those who represent them in government, whose vote will determine the quality of the air or water. At any given time there is debate at both local and national levels on environmental issues whose outcome will affect you. In this exercise you will have the opportunity to influence the outcome of the debate.

Procedures

During the exercise you will do the following:

1. Research an environmental issue. The issue may be either local or national in scope. It is important, however, that you determine where the issue will be resolved (e.g., the state capitol or the U.S. House of Representatives).
2. Compose and send a letter to an elected official who is involved in the debate. For political reasons, the official should be the one who represents you; however, you may also wish to write to one who does not represent you.

Directions

In drafting your letter it will be helpful if you follow the steps outlined. A personal letter is usually the most effective way of contacting your legislator, whether at the local or national level.

1. Address it properly: know your legislator's full name and its correct spelling. For specific addresses see individual lists. Examples:

   U.S. Senator
   The Honorable (full name)
   United States Senator
   Address
   ***
   Dear Senator (last name):

   U.S. Representative
   The Honorable (full name)
   United States Representative
   Address
   ***
   Dear Congressman/woman (last name):

   State Senator
   The Honorable (full name)
   State Senator
   State Capitol
   (Your State Capitol)
   ***
   Dear Senator (last name):

   State Representative
   The Honorable (full name)
   State Representative
   State Capitol
   (Your State Capitol)
   ***
   Dear Representative (last name):

2. *Always include your name and address on the letter itself (printed or typed).* A letter cannot be answered if there is no return address or the signature is not legible.
3. *Use your own words.* Avoid form letters and petitions. They tend to be identified as organized pressure campaigns and are often answered with form replies. However, a petition does let the legislator know that the issue is of concern to a large number of people (addresses with zip codes should be given for each signature). One thoughtful, factual, well-reasoned letter carries more weight than 100 form letters or printed postcards.

4. *Time for the arrival of the letter.* Try to write to your legislator, and the chairperson of the committee dealing with a bill, while a bill is in committee and there is still time to take effective action. Sometimes a bill is out of committee, or has been passed, before a helpful, informative letter arrives that could have made a difference in the way the bill was written or in the final decision.

5. *Know what you are writing about.* Identify the bill or issue of concern to you. Thousands of bills and resolutions are introduced in each session. If you write about a bill, try to give the bill number or describe it by popular title, such as "Land-Use Bill" or "Air Pollution Control Bill."

6. *Be reasonably brief.* Many issues are complex, but a single page presenting your opinions, facts, arguments, or proposals as clearly as possible is preferred and welcomed by most legislators.

7. *Give reasons for your position.* Explain how the issue would affect you, your family, your business or profession, the community, or your state. If you have specialized knowledge, *share it with your legislator.* Concrete, expert arguments for or against a bill can be used by the legislator in determining the final outcome of a bill.

8. *Be constructive.* If a bill deals with a problem you admit exists but you believe the bill is the wrong approach, explain what you believe to be the right approach.

9. *Groups and individuals should determine their priority concerns* and contact the legislator on those specific issues rather than on every issue. The "pen pal" who writes every few days on every conceivable subject tends to become a nuisance rather than an effective voice of concern.

10. *You may not always receive a long, detailed response* to your letter. Legislators are very busy and usually cannot respond with long, personal replies to each correspondent.

11. *Write a letter of appreciation* when you feel a legislator has done a good job. Legislators are human, too, and seldom receive "thank you" letters of encouragement.

Remember, on any one issue, even a few letters to one legislator can have an important impact. Sometimes a single letter from a new perspective or with a clear-cut, persuasive argument can be the decisive factor in a legislator's action.

Influencing Public Officials
Data Sheet

Name _____

Section _____

1. As a class project, identify a local environmental issue early in the semester or term and follow the actions of environmental groups addressing that issue. What strategies and tactics are used, and with what effects?

2. What means other than letter writing do you feel would be effective in influencing elected officials?

3. Describe the different ways in which you could participate in politics. Why do you think people are or are not participants? What factors do you feel influence participation or lack of participation in the political system?

Objectives

These will vary depending on the kind of field trip you are going to have. The instructor and students should both have a clear understanding of the objectives to be achieved during the trip.

Safety Considerations: These will vary, but you need to think about the possible safety problems associated with the trip.

Introduction

The purpose of a field trip is to provide a hands-on practical experience that helps people understand how theory and practice are interrelated. Every field trip should have a clearly established set of goals and objectives. These should be discussed and written down to focus the participants' attention on the important parts of the experience, so that they are not distracted by sights or activities that are peripheral to the trip's purpose. The participants may need to prepare for the trip by reading and discussing the roles each will take in meeting the objectives of the trip experience.

Procedures

1. Discuss and establish a clear set of objectives for the class to meet during the trip. Write these down and provide each student with a copy.
2. Assign responsibilities to class members.
3. Develop an equipment list if you plan to do an exercise that will require equipment.
4. Do a dry run if possible, so that you can work the bugs out of the procedure before you go on the trip.
5. Develop a report form, data sheet, or checklist, so that students can record their progress in meeting the objectives of the trip.
6. Plan a time for discussion and analysis of the data collected during the field trip, the errors that were made, and how effectively the learning experience met the planned objectives. If different groups had different assignments, they will need to report these results. If different groups were collecting the same kind of data, have them pool their data for analysis.
7. If the trip involves a visit to a place that has not been used previously, the instructor should visit the site to become familiar with the geography and other important features of the site.
8. Transportation needs to be arranged.
9. For legal liability reasons the school administration will require that certain practices and procedures be followed. Find out what these are.
10. If the field trip involves outdoor fieldwork, be sure to provide students with information about proper clothing, and be prepared for changes in the weather.
11. Each student should have a specific part to play in the plan for the day. If the class is divided into small groups with specific assignments, it is less likely that anyone will be left out.
12. Establish a clear set of acceptable and unacceptable behaviors. This is especially important when taking extended trips of a day or two. "Take nothing but pictures; leave nothing but footprints."
13. Use outside resource people if they are available.
14. If specimens are to be collected, establish guidelines about the number of individual organisms that should be collected for identification and study in the lab or as additions to an established collection. Avoid collecting large numbers of organisms that are just going to be thrown out when you get back to the lab.

A Checklist

1. Acquaint yourself with antipollution ordinances and make sure you abide by them. When you see a flagrant violation, report it to the proper authorities.
2. Don't burn leaves or trash. Better to start a compost heap and return the nutrients to the soil. Remove weeds in the lawn by hand rather than applying a herbicide.
3. Use insecticides sparingly and only when absolutely necessary. If you must use them, follow directions carefully.
4. Plant trees and shrubs. They absorb carbon dioxide, produce oxygen, help purify the air, and prevent soil erosion.
5. Use a hand mower if your lawn is small. Keep gasoline-powered tools in top operating condition to minimize noise and exhaust fumes.
6. Be careful with matches around wooded or grassy areas. Forest and grass fires cause air and water pollution.
7. When building a house, be sure it is well insulated and tree-shaded. This will minimize fuel consumption in winter and air-conditioning loads in summer.
8. In winter turn your thermostat down a few degrees. Have your home heating system checked annually or any time it appears to be operating inefficiently.
9. If you live in the city don't litter the sidewalks. Use litter baskets—and curb your dog.
10. Encourage and support your sanitation department when it seeks more modern and efficient collection and disposal equipment.
11. Start a campaign to save newspapers, cans, and bottles for collection and recycling where facilities are available.
12. Never flush away what you can put in your garbage pail. Organic materials, such as cooking fat, clog plumbing and septic tanks, causing sewage overflow.
13. Measure detergents carefully, using only enough to get your clothes clean. Try to run your dishwasher only once a day or less, depending upon the size of your family.
14. Don't use heavy electrical appliances, such as washers and dryers, during the hours when the electrical load is at its peak, usually 5–7 P.M. The strain at the local generating station may contribute to air pollution. Install low-wattage bulbs in lamps not used for reading. Turn out lights not being used to conserve power.
15. Don't drive a car when you don't have to. Walk, bicycle, or use mass transportation if possible. When you do drive, avoid quick starts and stops. Don't leave the engine running while parked. Car exhaust is a pollutant.
16. Make sure your car is equipped with required antipollution devices and have them checked regularly. If you buy a new car, read the instructions in your owner's manual regarding maintenance and upkeep of these devices. Match horsepower ratings to your needs. Don't buy a high-horsepower car if you don't need it.
17. Burn a fuel rated most efficient for your engine in terms of the reduction of emissions.
18. Get an engine tune-up every 10,000 kilometers or at least once a year. Be sure to change oil and air filters regularly.
19. Carry a litterbag in your car—and in your boat. Bring the bag back with you and dispose of it properly at the end of your trip.
20. Help reduce noise pollution. Don't use your horn unless safety dictates. Keep your muffler and tailpipe repaired.
21. Focus attention on the litter problem and efforts to combat it. Recommend that the student government set up good housekeeping rules for schoolwide use and assume responsibility for enforcement of them. Help create a school environmental improvement committee that includes both students and faculty. Run articles in the school newspaper.
22. Promote a local "cleanup" campaign. Business and service groups will often join in a task force to help spruce up a low-income neighborhood or public facilities such as railroad stations and overpasses.
23. Attend meetings of your local government and ask officials about their plans to control pollution. Often officials are responsive to visits of this kind.
24. Organize a community conference to discuss positive approaches to pollution control. Invite public officials, industry and labor representatives, other interested groups, and individual citizens. Get all the facts, and then make sure appropriate action programs are initiated.
25. Survey community opinion to determine how much support there is for clean air, water, and land programs. An effective community action program must know the extent of its support and the nature of its opposition.
26. Take an interest in local budget and zoning matters; attend public hearings, and be prepared to discuss their effect on your community. Try to work out problems cooperatively on the local level. In most cases everyone is looking for a solution.
27. Urge officials to provide better "street furniture." Benches, bus shelters, lampposts, street signs, and especially trash baskets all have a role to play in the elimination of visual blight. A street without trash baskets, for example, is frequently littered.
28. Urge public officials to adopt a sensible ordinance to govern the installation of commercial and industrial signs.
29. Seek homebuilders' assistance in developing a program to leave as many trees as possible on a subdivision when they develop it.

30. Work with garden clubs and other interested groups to landscape parks, malls, shopping centers, and similar areas.
31. Enlist the help of your local newspaper, radio station, and television station in pollution-control efforts. State your purpose clearly and forcefully to elicit strong editorial support. Be sure all the facts are presented. Avoid one-sided, prejudiced statements.
32. Ask national organizations and corporations for information and assistance. Many of them have experience in local environmental improvement programs and are willing to help.
33. Examine your workplace for pollution problems, and take the necessary steps to arrest them. What else can you add to this list?
34.
35.
36.

Environmental Plans

EPA work force in 1970: 5,500
EPA work force in 1990: 17,170
EPA budget in fiscal year (FY) 1971: $1,289,000,000
EPA budget in FY 1990: $5,145,000,000
Major general environmental statutes EPA administers: 11
Pages of EPA statutes (1989): 670
Total number of EPA regulations in 1989: 9,000
Times EPA officials testified before congressional oversight committees in 1989: 168
Letters to EPA's administrator and deputy administrator in 1989: 49,052

Environmental Concerns

Increase in U.S. population from 1970 to 1990: 48 million
Present ratio of increase in cars to the increase in U.S. population: 2:1
Estimated percent increase in ambient carbon monoxide by 1990 if 1970 emissions controls had remained unchanged: 140
Number of ozone molecules one chlorine atom can destroy: 10,000
Global carbon dioxide concentration in 1986 (in ppm): 346
Lung cancer deaths each year EPA estimates are due to radon: 5,000–20,000
Number of times the Superdome in New Orleans could be filled with the hazardous waste produced annually in the United States: 1,500
Superfund sites placed on the National Priorities List (NPL) by end of 1982: 418
Sites on the NPL by end of 1990: 1,207
Superfund sites with clean-up work completed to date: 52
Percentage of Americans who rely on groundwater as drinking water: 50
Estimated number of underground storage tanks leaking or potentially leaking and contributing to groundwater contamination: 400,000
Average level of DDT found in humans in 1970 (in ppm): 8.0
Average level of DDT found in humans in 1983 (in ppm): 2.0

The chi-square test is a statistical method of determining if the results of an experiment are close enough to what was expected to be considered valid. A chi-square test answers the question, Was it chance that this deviation from the expected results occurred, or is something else going on here? If the results of an experiment always occur according to expectations, there is no deviation from what was expected. The deviation from expected is zero, and the results are considered valid. However, usually the results of experiments do not turn out perfectly. They are not *exactly* what was expected. Scientists agree that chance events can cause results to vary from the expected outcome. Deviation from the expected results could also be nonrandom and be caused by outside influences that the experimenter did not anticipate. How can chance variation be differentiated from variation caused by outside influences? It is important to have guidelines about how much deviation from the expected results is acceptable.

If the deviation from expected results is very small and the number of events is very large, random variation is probably the cause of the deviation. However, if the deviation is large or the number of events is small, statistical tests are needed to separate random variation from variation having other causes.

For example, we would predict that if we flipped a coin 100 times, we should get 50 heads and 50 tails. If we actually did the experiment and got 50 heads and 50 tails, there would be no deviation from expected and we would conclude that our prediction was correct. However, what if the results of flipping the coin gave 45 heads and 55 tails? Is the deviation from the expected a random variation or is the coin biased in favor of tails? The chi-square test helps us answer that question. Table A.1 shows how the chi-square value is determined.

$$X^2 = \Sigma \frac{(O-E)^2}{E}$$

Table A.1 Determining the chi-square value.

Classes	Observed (O)	Expected (E)	$O-E$	$(O-E)^2$	$\dfrac{(O-E)^2}{E}$
Heads	45	50	−5	25	0.5
Tails	55	50	5	25	0.5

Chi-square value = _____ 1.0 _____

The chi-square value is 1.0. To interpret what this value means, we must consult table A.2. In order to use this table we must first determine the degrees of freedom. The degrees of freedom is equal to the number of classes minus one. In this case there are two classes (heads or tails). The degrees of freedom is one (2 − 1 = 1). Look along the row for one degree of freedom until you find the chi-square value that most closely approximates 1.0 (in this case it is 1.074), then look down the column to the last row labeled "confidence level" and you will find the number 0.30. This number means that the deviation from expected obtained should be expected to occur 30% of the time simply as a result of random variation. This deviation from expected is within the acceptable range and our original prediction of 50/50 heads and tails is still acceptable.

Table A.2 Chi-square (X^2) values.

Degrees of freedom								
1	0.0002	0.004	0.455	1.074	1.642	2.706	3.841	6.635
2	0.020	0.103	1.386	2.408	3.219	4.605	5.991	9.210
3	0.115	0.352	2.366	3.665	4.642	6.251	7.815	11.345
Confidence level	0.99	0.95	0.50	0.30	0.20	0.10	0.05	0.01

However, what if we flipped a coin 100 times and got 60 heads and 40 tails? Is this acceptable? As shown in table A.3, we can use the chi-square test to help us decide.

Table A.3

Classes	Observed (O)	Expected (E)	$O–E$	$(O–E)^2$	$\dfrac{(O–E)^2}{E}$
Heads	60	50	10	100	2
Tails	40	50	−10	100	2

Chi-square value = _____4_____

The chi-square value is 4. The number of classes is 2. There is one degree of freedom. Look along the row for one degree of freedom until you find the chi-square value that most closely approximates 4, then look down the column to the "confidence level" row. From the table, we can see that the deviation from expected we obtained would occur less than 5% of the time as a result of random events. This is very low probability. Generally, if chi-square tests result in probabilities of 5% or less, the deviation is considered to be the result of something other than random variation. The prediction that we should get 50/50 heads and tails is wrong. Perhaps the coin is weighted in such a way that heads is more common than tails.

What if we flipped the coin only ten times and got six heads and four tails?

Table A.4

Classes	Observed (O)	Expected (E)	$O–E$	$(O–E)^2$	$\dfrac{(O–E)^2}{E}$
Heads	6	5	1	1	0.2
Tails	4	5	−1	1	0.2

Chi-square value = _____0.4_____

The chi-square value is 0.4. The number of classes is 2. There is one degree of freedom. Look along the row for one degree of freedom until you find the chi-square value that most closely approximates 0.4; then look down the column to the "confidence level" row. From the table, we can see that the deviation we obtained would be expected to occur 50% of the time as a result of random variation. In this situation there is no significant deviation from what was expected and our original prediction is supported. This indicates the importance of the size of the sample. The previous case of 60 heads and 40 tails was significantly different from the expected ratio of 50/50. However, 6 heads and 4 tails is not significantly different from the expected ratio of 5/5.